Lecture Notes in Economics and Mathematical Systems

Managing Editors: M. Beckmann and H. P. Künzi

Economics

126

Energy, Regional Science and Public Policy

Proceedings of the International Conference on Regional Science, Energy and Environment I
Louvain, May 1975

Edited by M. Chatterji and P. Van Rompuy

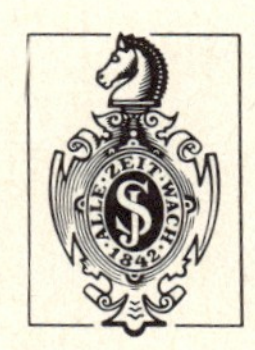

Springer-Verlag
Berlin · Heidelberg · New York 1976

Editors
Manas Chatterji
School of Management
State University of New York
Binghamton, N. Y. 13901/USA

Paul Van Rombury
Department of Economics
Centrum voor Economische Studiën
(Katholieke Universiteit Louvain)
Van Evenstraat 2B
3000 Louvain/Belgium

Library of Congress Cataloging in Publication Data

International Conference on Regional Science, Energy, and Environment, Louvain, 1975.
Energy, regional science, and public policy.

(Proceedings of the International Conference on Regional Science, Energy, and Environment, Louvain, May 1975 ; v. 1) (Economics) (Lecture notes in economics and mathematical systems ; 126)
1. Energy policy--Congresses. 2. Regional planning--Congresses. 3. Regional economics--Congresses. I. Chatterji, Manas, 1937- II. Rompuy, Paul van. III. Title. IV. Series: Economics (Berlin, New York) V. Series: Lecture notes in economics and mathematical systems ; 126.
HT391.I475 1975 vol. 1 [HD9502.A2] 309.2'5'08s
76-10466

AMS Subject Classifications (1970): 90A15, 90B05, 90B20, 90B99

ISBN 3-540-07692-1 Springer-Verlag Berlin · Heidelberg · New York
ISBN 0-387-07692-1 Springer-Verlag New York · Heidelberg · Berlin

Printed in Germany
Printing and binding: Beltz Offsetdruck, Hemsbach/Bergstr.

CONTENTS

PREFACE V

CONTRIBUTORS VII

WORLD ENERGY SITUATION 1
Manas Chatterji

FORECASTING ALTERNATIVE REGIONAL ELECTRIC REQUIREMENTS AND ENVIRONMENTAL IMPACTS FOR MARYLAND, 1970-1990 32
John H. Cumberland, William Donnelly, Charles S. Gibson, Jr., and Charles E. Olson

THE LOCATION OF ENERGY-RELATED INDUSTRIES IN WESTERN EUROPE 58
W. T. Molle and Henk Breimer

REGIONAL BALANCED GROWTH IN ITALY AND THE INCREASE IN OIL PRICES 78
Murray Brown, Maurizio Di Palma and Umberto Triulzi

A DYNAMIC REGIONAL ANALYSIS OF FACTORS AFFECTING THE ELECTRICAL ENERGY SECTOR IN THE U.S. 90
Rajendra K. Pachauri

PERSPECTIVES ON SHORT TERM ENERGY SHORTAGES IN THE NETHERLANDS 104
P.J.J. Lesuis and F. Muller

A LINEAR PROGRAMMING MODEL FOR DETERMINING AN OPTIMAL REGIONAL DISTRIBUTION OF PETROLEUM PRODUCTS 119
Craig L. Moore and Andris A. Zoltners

THE WISCONSIN REGIONAL ENERGY MODEL: A DYNAMIC APPROACH TO REGIONAL ENERGY ANALYSIS 136
Wesley K. Foell, John W. Mitchell and James L. Pappas

ESTIMATING THE REGIONAL IMPACTS OF ENERGY SHORTAGES 178
William Donnelly and Ali M. Parhizgari

TECHNOLOGICAL ABATEMENT VS. LOCATIONAL ADJUSTMENT: A TIME SPACE DILEMMA 194
Manoucher Parvin and Gus W. Grammas

COMMENTS ON PARVIN AND GRAMMAS' PAPER 217
S. De Kock

FROM WAGES TO BADLY CIRCULATING RENT 220
A.R.G. Heesterman

Contents (continued)

COMMODITY MODELING APPROACHES TO RESOURCES, ENERGY AND REGIONAL PLANNING 250
Walter C. Labys

DETERMINATION OF SOCIAL COSTS OF ENVIRONMENTAL DAMAGE 262
A. P. Mastenbroek and P. Nijkamp

THE RISKS AND BENEFITS FROM LEAD IN GASOLINE: EFFECTS ON ENERGY USE AND ENVIRONMENT 291
Robert W. Resek and George Provenzano

COMMENTS ON RESEK-PROVENZANO PAPER 303
Peter M. Meier

TECHNICAL PROGRESS IN AGRICULTURE AND ITS IMPLICATIONS FOR ENERGY USE 307
L. von Bremen

URBAN DESIGN SHAPING THE ENVIRONMENT -- A NOTE 315
Vladimir Music

PREFACE

For the last three decades, space has become a significant dimension in social science analysis. In many developed countries, economic growth is slowing down, and in some cases restrained, due to environmental considerations, and the real question is the optimum spread of development over space rather than the growth over time. In the developing countries, limited and uneven distribution of population and resources, and the existence of heterogenous groups, highlighted the need of balanced regional development. The energy crisis and the realization that energy resources are very limited and unequally distributed have further emphasized its importance. The expected impact and relocation due to energy shortages will have a crucial spatial dimension since manufacturing and service activities dependent on energy are concentrated in a few metropolitan regions connected by transportation, communication and cultural factors. Regional environment is also dependent on the spatial juxtaposition of activities and energy use.

The papers included in this volume address some of these considerations. They were presented in the International Conference on Regional Science, Energy and Environment, held at Katholieke Universiteit te Leuven, Belgium, on May 1975. A second volume titled Environment, Regional Science and Interregional Modeling published also by Springer-Verlag contains research papers related to environment and space. This book does not concern all facets of the energy situation. In fact, contributions on major oil producing countries, U.S.S.R., East European and developing countries are not represented. Again many factors like politics of energy, price, petro-dollar, foreign exchange, sociological impact, etc. are not analysed. Hopefully, on another occasion these items could be specially covered. We are grateful to the participants of the conference, administration and economics faculty of Katholieke Universiteit te Leuven, several banking and governmental organizations in Belgium, for making the conference successful. Particular reference should be made of R. Donckels, F. Moulaert, A. Tejano, L. Van Overbeke and others in the Centrum voor Economische Studien at Katholieke Universiteit te Leuven. We also wish to thank Mr. Amrik Sethi for helping us with the editorial work of this volume.

Manas Chatterji
State University of New York at
Binghamton, U.S.A.

Paul Van Rompuy
Katholieke Universiteit te Leuven, Belgium

February 1976

CONTRIBUTORS

HENK BREIMER	The Netherlands Economic Institute Rotterdam, The Netherlands
MURRAY BROWN	Department of Economics State University of New York at Buffalo, N.Y. U.S.A.
MANAS CHATTERJI	School of Management State University of New York at Binghamton, N.Y. U.S.A.
JOHN H. CUMBERLAND	Department of Economics University of Maryland College Park, Md., U.S.A.
S. DE KOCK	Centrum Voor Economische Studien Katholieke Universiteit te Leuven Leuven, Belgium
M. DI PALMA	Centro di Studi e Piani Economici Rome, Italy
WILLIAM DONNELEY	Federal Energy Administration U. S. Government Washington, D.C., U.S.A.
W. K. FOELL	Institute of Environmental Studies University of Wisconsin Madison, Wis., U.S.A.
CHARLES S. GIBSON	Department of Economics University of Maryland College Park, Md., U.S.A.
G. GRAMMAS	Graduate School of Business Columbia University New York, N.Y., U.S.A.
A.R.G. HEESTERMAN	Department of Economics University of Birmingham Birmingham, England
W. C. LABYS	Graduate Institute of International Studies Geneva, Switzerland
P.J.J. LESUIS	Erasumus Universiteit Rotterdam, The Netherlands
A. P. MASTENBROEK	Erasmus University Rotterdam, The Netherlands
P. MEIER	Brookhaven National Laboratory Upton, L.I., N.Y., U.S.A.
J. W. MITCHELL	Institute of Environmental Studies University of Wisconsin Madison, Wisconsin, U.S.A.

Contributors (continued)

W.T.M. MOLLE — Netherlands Economic Institute
Rotterdam, The Netherlands

C. L. MOORE — School of Business Administration
University of Massachusetts
Amherst, Mass., U.S.A.

F. MULLER — Erasumus Universiteit
Rotterdam, The Netherlands

V. MUSIC — Urbanisation Institute
Ljubljana, Jugoslavia

P. NIJKAMP — Vrije Universiteit
Amsterdam, The Netherlands

CHARLES E. OLSON — Department of Economics
University of Maryland
College Park, Md., U.S.A.

R. K. PACHAURI — Administrative Staff College
Hyderabad, A.P., India

J. PAPPAS — Institute of Environmental Studies
University of Wisconsin
Madison, Wis., U.S.A.

A. M. PARHIZGARI — Federal Energy Administration
U. S. Government
Washington, D.C., U.S.A.

M. FARVIN — Department of Economics
Fordham University
New York, N.Y., U.S.A.

G. PROVENZANO — Department of Economics
University of Illinois
Urbana, Ill., U.S.A.

R. RESEK — Department of Economics
University of Illinois
Urbana, Ill., U.S.A.

U. TRIULZI — Centro di Studi e Piani Economici
Rome, Italy

L. VON BREMEN — Institut fur landwirtschaftliche Marktforschung
Braunschweig, West Germany

A. ZOLTNERS — School of Business Administration
University of Massachusetts
Amherst, Mass., U.S.A.

WORLD ENERGY SITUATION

Manas Chatterji

With the spread of urbanisation and industrialisation, orientation of human activity is changing from production to service industries. A high standard of living is usually associated with high degree of service activities. However, the crucial importance of resources and their management has not diminished. As the world population grows and the standard of living of the developing countries accelerates as well, the demand on resources will increase. With the operations of multinational corporations for exploitation of resources for meeting the raw materials need of the developed countries, to ship food grain abroad to stop hunger and starvation and to continue prosperity at home, the pressure on resources would remain strong in the developed countries and its different regions. With the passage of time, political systems are becoming more and more decentralised leading to local decision making. Thus, resources management is and will be a top priority on local, regional, national and international administration.

It is true that resources and material goods alone cannot bring happiness. But a minimum amount of resources is prerequisite for basic necessities and comfort. There are many ways resources can be exploited and used. They all concern the economic, political, moral psychological and military fabric of the society in the context of a geographical space. This space has different characteristics namely terrain, location, climate, vegetation, wildlife, minerals, etc. Although their locations are more or less known, the spatial configurations of the most efficient productive process is by no means defined. Dynamics of space in the framework of resource endowments, new discoveries, tehcnological changes, externatlities, etc., are far from well understood at present time. Resource mobilisation goes far beyond the political boundary and it has entangled countries in fierce power game. It also affects the political system of a country through control of prices of raw materials, etc. In fact, even for a free enterprise countrly like the United States, a considerable amount of activity is regulated. This brings us to the question of individual or corporation interest against that of society as a whole. There are many who think resources should be exploited to a point where it accrues profit to the individual or corporation without regard to the needs of the future generation. On the other hand, there are some who believe that we have enough and that we should preserve the resources for the future. There is the third group who argues against waste and they propagate for wise use of resources. The problem is to define a wise decision. However, it can be said that in many countries the concept of collective interest as compared to individual profit maximization motive is becoming the guiding force of public policy.

Resources can be either tangible or intangible. Some of the tangible resources are:

1) Energy
2) Forests
3) Air
4) Water
5) Soil
6) Labor, etc.

Intangible resources include technology, moral and ethical standards, cultural and religious background. The amount of tangible resources can be measured. Although this is not so in the case of intangible resources, they are important as well and different measurement techniques have been devised in social and behavioral sciences in recent years.

Here we are interested in energy resources. It is true that we cannot study energy in isolation since it is interrelated with other resources. This interrelation between energy and other resources coupled with interdependence amongst sectors and regions is crucial in economic analysis.

The concept of energy is as old as human beings themselves and it is inherent in many religious and philosophical writings. It is the basic concept of physical science in areas of classical mechanics, thermodynamics, engines, electrochemistry, magnetic theory, theory of relativity and quantum mechanics, etc.

Let us now define the measuring units of energy. One unit of 'erg' (= 1 gram - cm^2/sec^2) is defined as the work done by a force of 1 dyne (= 1 gram - cm/sec^2) during a displacement of 1 cm. This is known as c.g.s (centimeter - gram-second) system. If we use kilogram-meter-second (kg. m. s) unit then one 'newton' is defined as the unit of force with 1 newton = 10^5 dyne. 15° Calorie is the quantity of heat or energy required to raise the temperature of 1 gram of liquid water from 14.5° to 15.5° C. 60° Btu (British thermal unit) is the energy required to raise the temperature of one pound of mass of water from 59.5° to 60.5°.

The 15° Calorie and 60° Btu are usually replaced by the international table calorie (I.T. cal) or Btu. One Btu is thus the amount of energy needed to raise the temperature of 1 lb. of water by 1 fahrenheit degree. For example, we require about 150 million Btu to heat an average house for a year and 100 million Btu to run an average car. One barrel (bbl.) of petroleum contains 49 gallons which can produce 5.8 million Btu and one metric ton (mt) or 2200 lb. of coal has about 25 million Btu of equivalent energy. One standard cubic feet (SCF) of natural gas has about 1000 Btu whereas 1 cord (128 c.f.) of wood has about 20 million Btu. It is to be remembered that in the process of transformation of energy some energy is lost. For example, in the case of nuclear reactor the efficiency varies between 35-65%. Sometimes 10^5 = 1 quadrillion is used as a unit to measure energy consumption. For example, in 1973, the total consumption of energy in the U.S. was 73 quadrillion of Btu which is about 30% of the world consumption when its population was 6% of world population. The supply from internal sources was 62 quadrillion. Its gasoline equivalent was 7 gallons/person/day equally divided between electricity generation, transportation, manufacturing, commercial and residential sector. It is estimated that per capita consumption of energy in the U.S. is about 11.7 KW (thermal equivalent) per year, and this will increase to 17.8 in 2000. The corresponding figures of the world are 2, 10 respectively.

There are many sources of energy. Some of them are:

1) Natural gas (mostly methane)

2) Natural gas liquids (NGL) obtained during the production of natural gas.

3) Liquified petroleum (LPG) attained as a mixture of propane and butane

4) Liquid crude oil obtained from underground reservoirs by means of oil wells. It is a heavy viscous material with green-red to black color. It is the resultant of partial oxidation of vegetables and animal debris buried in the sedimentary rocks.

5) Tar sands which is a viscous organic material found in sand, mostly silt and clay.

6) Oil shale which is a waxy solid hydrocarbon.

7) Coal which can be smokeless anthracite or soft bituminous or lignite.

8) Nuclear fuels such as uranium where 1 K.g. of U^{235} material has energy equivalent of 1.34×10^6 bbl. of petroleum. The energy potential of fusion energy sources is immense.

9) Solar energy

10) Hydrothermal energy

11) Geothermal energy

12) Tidal energy

13) Wind energy

14) Other sources.

Due to time and space limitation we shall restrict ourselves to a few sources namely coal, oil, natural gas and nuclear energy, although some summary remarks will be made regarding other sources.

One important point to remember is that the definition of a resource changes over time and space. Technology is fast changing and world economic conditions, particularly prices, cost of production, subjective judgment of experts, all play an important role in the definition of a resource. There are also some problems in terminology. To some people the term petroleum denotes both liquid and gaseous hydrocarbons. However, in general, it refers to liquids alone and natural gas refers to gaseous hydrocarbons. Again a clear distinction has to be made with the words resources and reserves. If we follow the U.S. Geological Survey (USGS) definitions, resources refer to concentration of naturally occurring solids, liquids, and gaseous material in or on the earth's crust discovered or surmised to exist in such a form that economic extraction is currently or potentially feasible at higher future prices.[1]

The availability of resources depends on the degree of certainty about the quantity and quality of resources and economic possibility of extraction. In the extreme the resources may have been already identified or undiscovered. Economic possibility may refer to recoverable, paramarginal or submarginal. Identified resources are those whose location, quantity and quality are known by geological and engineering measurements.

The term "reserve" refers to those resources which can be recovered with existing technology and present economic conditions. So reserves are recoverable identified resources.

Identified resources could be _measured_, _indicated_ and _inferred_. _Measured_ resources are those whose (quality and quantity) have been estimated within a 20% margin of error. The term _indicated_ is applied to resources whose quality and quantity has been estimated partly by some sample analyses and reasonable geological projections. The term _inferred_ refers to unexplored but identified resources.

Undiscovered resources are those surmised to exist on the basis of geological evidence and theoretical considerations.

Paramarginal resources are those which are recoverable with prices as much as 1.5 times those prevailing now. Legal and political difficulties are preventing its exploitation.

Submarginal resources are those which cannot be recovered under the existing technological or economic conditions because of size and location (more than 1.5 times the price required).

These terms can refer both to identified and undiscovered resources. Of course, some resources which are termed as submarginal could become paramarginal

---

[1] U.S. Department of the Interior, "New Mineral Resource Technology Adapted", April 15, 1974 (press release).

if prices become sufficiently high. Similarly paramarginal could become part of recoverable resources. According to USGS, the estimates of identified recoverable resources has 20-50% error but if it is measured the error is less than 20%.

Undiscovered recoverable require continuing effort in exploration research and economic incentive.[2]

Let us next briefly discuss the energy situation in the world as a whole and some selected countries. Table 1 gives some data about past and projected demand and supply of energy sources for some geographical regions of the world.

TABLE 1

World Energy (10^6 bbl/d)

	1971	1976	1990
A. Demand			
U.S.A.	15.1	22.9	35.0
W. Europe	13.2	19.8	36.0
Japan	4.5	7.9	27.0
R.W.	8.5	11.5	19.6
	41.3	62.1	117.6
B. Supply			
U.S.A.	11.6	10.0	14.5
Middle E.	15.0	30.9	57.6
Africa	5.8	8.5	17.0
R.W.	8.9	12.7	28.5
	41.3	62.1	117.6

It is imperative that a thorough historical study be done about the world energy supply and demand and correlate them with other activities. In the absence of such a study we can say that study of the past trend for noncommunist countries (U.S., W. Europe) brings forward the following points.

1) Nuclear energy is relatively new in the field. It has been said that it takes about 10 years for a nuclear power station to supply energy from its inception.

2) In almost all countries since 1955, the growth rate of consumption of petroleum and natural gas have been more than in coal or hydroelectric power.

3) Nuclear power growth rate is highest since 1970.

---

[2] The definitions are based on V.E. McKelvey, et al.; "Minerals Resource Estimates and Public Policy," American Scientist, 60, pp. 32-40, 1972.

4) There has been a significant increase in the demand of natural gas in W. Europe due to recent discovery in the North Sea and the supply of gas from U.S.S.R.

5) Although Japan is not an oil producing country, the import of oil in Japan is increasing at about 11% annual rate as contrasted about 5% growth rate in the U.S. and W. Europe.

6) Historically speaking, the U.S. had a disproportionately higher share compared to other countries of the world. But its share of world consumption is decreasing.

We next propose to examine the energy situation in some specific countries.

United States

The history of economic development of the U.S. is a direct function of its energy consumption. During the period 1967-73, energy consumption has increased more than the growth in gross national product. From 1950 to 1973 the consumption grew at about 3.5 per cent where as the domestic production grew about 3 per cent per year. After 1965 the growth rate in consumption rose to about 5% per year but since 1970 there was no growth in domestic production at all. Tables 2-7 give some information about energy reserve situations in the U.S. (Source: Energy Policy Project of the Ford Foundation, A Time to Choose, Appendix D.)

TABLE 2

Coal as of 1972
(Trillions of short tons)

Total estimated coal resources remaining (0-3000 feet)	1.58
Total estimated undiscovered coal	1.64
Total coal resources (0-6000)	3.22

TABLE 3

Recoverable Petroleum as of February 1974
(billions of barrels)

	Measured	Indicated & inferred	Undiscovered Recoverable resources
Total onshore	41	22-39	135-270
Total offshore	8	3-6	65-130
TOTAL	49	25-45	200-400

TABLE 4

Petroleum as of December 1970 (Including Alaska)
(billions of barrels)

	Identified (remaining)	Undiscovered
Recoverable	50	150-450
Paramarginal & submarginal	290	280-2100

TABLE 5

Recoverable Natural gas as of February 1974
(trillions cubic feet)

	Reserves		
	Measured	Indicated & Inferred	Undiscovered Reserve
Total onshore	218	107-205	605-1210
Total offshore	48	23-45	395-790
	266	130-250	1000-2000

TABLE 6

Natural Gas as of December 1970 (Including Alaska)
trillion cubic feet)

	Identified	Undiscovered
Recoverable	290	1200-2100
Paramarginal & Submarginal	170	4000

TABLE 7

Oil Shale as of 1972
(billions of barrels by oil yield)

	Identified	Undiscovered
25-100 gallons/ton	418	900
10-25/ton gallons/ton	1600	25,000
5-10 gallons/ton	2200	138,000

Data given in the above tables should not be taken as definitive. Unfortunately the U.S. government itself compiles very little information and mostly depends on private agencies.

Wood burning was the most important source of energy in the initial years. In 1850, its share was about 90% and the share of coal was 10%. Coal's share gradually increased to 28% in 1870 and in 1884 coal became more important than wood burning. The growth rate for oil and natural gas liquids was similar and its share outmatches that of wood burning in 1890. The 1970 demand and projected value for 1985 are given in Table 8.

TABLE 8

U.S. Energy Consumption (10^6 bbl/d equivalent)

	1970	1985 (estimate)
gas	11.00	14.1
coal	6.50	7.2
oil	14.70	33.3
hydroelectric	1.25	1.4
nuclear	.10	10.0
wood	.38	.3
Total	33.93	66.3
Per Capita/year (Equivalent Thermal Energy)	60	100

Source: S. Field, SNG and the 1985 Energy Picture, Stanford University Press, 1972

It is to be noted that in 1970 the consumption of wood is 4 times more than from nuclear power. The projected value for oil in 1985 is much too high particularly in view of the fact that the total oil consumption has actually declined in recent months. A more conservative estimate is given in Table 9.

TABLE 9

Alternative Estimate of U.S. Energy Consumption
(10^6 bbl/d equivalent)

Year	
1970	32.6
1971	33.2
1972	34.5
1973	36.3
1974	38.1
1975	39.9
1976	41.7
1977	43.5
1978	45.4

Source: S. Clark, World Energy, Stanford Research Institute

It is interesting to note that 1973 oil consumption projection was not realised due to oil embargo.

If we consider the consumption of natural gas and (crude oil & NGL) and assume 6% growth annually, then the entire world's output of the reserves of the world will be used by 2020 and by 1990 all the resources in the U.S. will be used up.

U.S. consumption rates are so high that consideration of all policies lead to the conclusion that even with the assumption of high resource existence and import possibilities and synthetic fuel conversion, the time difference will not be more than 10-15 years.

It is to be noted here that the projections are very much dependent on the political situation. For example, the projection made by the National Petroleum Council on the basis of 1972 consumption became too exaggerated due to the acceleration of price rise and composition of the demand. The contribution of oil consumption to total energy (bbl/d/person) in the U.S. as compared to other countries is given in Table 10.

It is to be noted in this connection that although the per capita oil consumption in Sweden is more than in the U.S., the total energy consumption equivalent is only 82%.

TABLE 10

The Contribution of Oil Consumption to Total Energy
(bbl/d/person)

U.S.A.	.155
Canada	.128
U.K.	.075
France	.059
Japan	.055
R.W.	.012

Coal in the U.S. is the most abundant domestic fuel. Although coal can be used as a substitute in many cases, this substitution is by no means universal and the technological problems involved are yet to be solved. Another basic problem is environmental. Extensive coal exploitation will also lead to problems of land use and transportation. Environmental considerations are also crucial in the case of offshore exploitation. However, it is expected that a considerable amount of coal is going to be used in the future although as seen from the following table, the bituminous coal production has not changed drastically in different countries.

TABLE 11

Bituminous Coal Production
(10^6 mt)

	1970	1971
U.S.	541	495
U.S.S.R.	432	480
U.K.	144	147
Poland	140	145
W. Germany	111	111

Natural gas is a clean fuel, easily transportable and low pollutant energy source. Since this source of energy has been used very rapidly and the price is relatively low, it will disappear from the energy scene very soon. The total demand for natural gas in 1970 was about 22.7 (10^{12} SCF), and the percentages in different sectors and regions were the regional breakdowns.

TABLE 12

Percentage of Natural Gas Consumption in Different Sectors

Residential & commercial	32.7
Industrial	44.5
Electric Power	18.2
Transportation	2.3
Manufacture of Chemicals	2.3
	100.0

TABLE 13

Regional Breakdown for Natural Gas Consumption

	Percentage
West South Central	36.6
East North Central	17.9
W. North Central	9.1
Pacific	11.2
Middle	8.4
New England	1.1
Remaining	
R.W.	17.9
	100.0

Texas was the heavy exporter and the Pacific area was the prime importer. The demand is related to availability and not with the population and other socio-economic variables. The prices charged to residential use are the highest and power plants the lowest. But the cost to supply the residential areas is much higher. The projected gas supplies are shown in Table 14.

TABLE 14

Total Projected Gas (10^{12} SCF)

	Supply	Residential & Commercial	Industry & Electricity
1970	22.5	7.5	15.2
1975	26.0	8.8	17.2
1980	25.7	10.8	14.9
1985	29.2	13.7	15.5
1990	32.0	15.9	16.1

In 1973 the source of energy supplies was as follows:

Coal	15%
Natural gas	30%
Oil	50%
other	5%
	100

and the demand was

Industrial	43%
Residential	19%
Commercial	14%
Transportation	24%
	100

In 1972, when we consider a barrel of refined oil, 38% was (gasoline) used in transportation (cars, trucks and buses); 18% (distillate fuel oil) for home heating, deisel fuel for trucks, etc.; 15% for residual fuel oil for electrical power plants and other industrial use, natural gas liquid for industrial and farm uses (14%) and the remaining 15% for other uses. For example, in 1972, out of total demand of 16.2 bbl/d for petroleum product, 6.3 bbl/d is for motor gasoline. It is anticipated that in 1978, out of 25.6 bbl/d, the demand for gasoline for cars will be for 8.5 bbl/d.

It is expected that the energy consumption by utilities to produce electricity will have a substántial growth from 25% in 1971 to 38% in 1990. During 1968, about 14% of the energy was consumed in the basic materials like iron and steel, aluminum, etc. Output of that above 7% was in iron and steel industry which used about 12 thousand (kwh_{th}) per ton. The lowest energy cost per ton of products was for sand and gravel and highest for rolled titanium. For aluminum it is about 67 thousand (kwh_t).

Energy consumption in transportation was 20.17 (10^{15} Btu), of which about 50% was in automobiles, trucks and buses. It has also been estimated that changing to returnable bottles and refilling them eight times, energy saving of about two-third can be accomplished per gallon of beverage. The ranking of primary energy consumers are:

primary metals	21%
chemicals	20%
petroleum	11%
food	5%
paper	5%

(out of total industrial energy consumption).

The chemical and allied products, primary metals, and petroleum and coal industries were the leading fuel users. The most fuel intensive manufacturing are stone, clay, and glass. They also had the highest fuel cost/$ value shipped. With respect to electricity, primary metals, chemicals, and allied products have the highest consumption, highest intensity, and cost/$ of product shipped.

Within the confines of the limited space here, it is not possible for us to discuss every aspect of the energy situation in the U.S. and the policy implication. Recently some of these problems have been dealt with in the Energy Policy Project supported by the Ford Founcation.[3] Three Scenarios have been considered. The first termed as historical growth (HG) is a situation where it is postulated that the future will be the replica of the past namely on growth rate of 3.4% per year in energy consumption. The long term implication of this hypothesis is that we are going to have smaller total output due to the projected decrease in labor force so that per capita income is not reduced. Output in transportation will increase most rapidly and servies least rapidly. Demand for travel will increase and due to more electricity demand pressure on primary energy sources will soar. Whereas the growth in agriculture will be slow due to low population, manufacturing will clearly follow the production growth. But the value added in manufacturing will rise slowly due to higher productivity. The total employment will increase by 1.7% but decline later due to low fertility rate. The prices of fuels will rise sharply not so much in the case of electricity, due to historical reasons. The main feature of historical growth scenario is the growth in electricity demand and in primary inputs like coal and nuclear power. Although the report found that maintenance of HG Scenario will not have devastating impact on the economy, such a situation is practicable. By the year 2000, the total energy consumption will be reduced by 60%. The drop in real GNP will be about 3.8% and there will not be any major increase in unemployment. This will result in an increase of 60% in the petroleum and about 50% in electricity prices.

The third scenario termed as zero energy growth (ZEG) hypothesize a tax system to discourage energy consumption and channel the tax revenue for social services. The result of such a policy in the year 2000 is not very much different from that obtained in Technical Rise Scenario.

In spite of the optimistic findings in this study, it can be safely concluded that the United States faces a serious energy crisis and unless the country reduces its consumption and waste, and adopts conservation measures, and unless a new breakthrough in technology is achieved, the country is in for big trouble ahead.

---

[3]Energy Policy Project of the Ford Foundation, A Time to Choose. Balinger publishing Co.

Japan

One of the major wonders in modern times is the economic development of Japan. With this development came the insatiable demand for energy with negligible internal supply.

Energy Production

The explosive growth of industry in the past twenty years has brought about a major expansion in the demand for power and fuels. With this has come an increasing concern by government and industry over growing shortages and rising costs of production. This increase in need for power has been running at a rate of ten to twenty per cent per year and government has supported industry in major expansion programs.

The role of energy in modern Japan has changed in recent years. Consumption of electricity and petroleum has expanded rapidly, while coal has changed very little. This follows the general world trend. The tremendous expansion in the use of petroleum has been of major significance to Japan. Japan is now one of the world's biggest markets for petroleum, and in fact it is their number one import.

Water power is one of Japan's most important natural resources. The mountainous terrain, abundant rainfall, and swift streams provide the power for one of the world's largest hydro-electric industries. Hydro stations are found throughout Japan, with the heaviest concentrations being in central and northern Honshu. The concentration in central Honshu is particularly good because of its location near the center of the industrial belt lying between Tokyo and Kobe. Thermal plants have been used during dry periods and the winter, but increasing demand has brought many changes to the electricity industry. In 1950, thermal power was only fifteen per cent of consumption. Today it is more than three times as much. Hydro plants are still being constructed, but costs are going up, and many of the good sites are already developed.

The rising cost of hydro-electric power, limits of coal supply, and increasing dependency on imported fuels, especially petroleum, gave incentives for development of atomic energy. At this point, Japan's future in atomic energy is unclear. But recent developments in the oil producing nations should make atomic power economically feasible. There are still, however, hard feelings and distrust among the population, since Japan was the first testing ground of the atomic bomb. Today Japan's atomic energy use is twelve times what it was five years ago.

Japan, like most of western Europe, is a major consumer of petroleum and, like western Europe, it depends on imports for practically all of its needs. Japan only produces about fourteen per cent of its petroleum consumption. Japan's total petroleum resources would only take care of the United States for a few days.

Most of Japan's production of oil comes from the northwest coast of Honshu between Niigata and Akita. Reports of offshore findings have caused some excitement, and rising costs of imports may make this economically feasible. But the outlook is not too good, because Japan's petroleum needs dwarf what they may be able to produce.

The United States was a major source of Japan's pre-war petroleum needs, but now the Middle East supplies the bulk of Japan's imports. One significant facet of Japan's energy resources is the Japanese concession in the offshore area of the neutral territory controlled by Kuwait and Saudi Arabia on the Persian Gulf. The Arabian Oil Company of Japan holds a forty-four and a half year concession in an area that gives promise of being one of the major new fields in the Middle East. The company has started production and was producing sixty million barrels per year in 1966. Some reports give very high expectations. These developments could put Japan into a much sounder position. Japanese business has followed its tradition of being very aggressive. In fact, its aggressive purchasing policy during the embargo was a major cause of high prices charged on oil. Japan's power needs are expected to double in the next five years, and thus we can see the overpowering need for Japan to continue to expand their energy resources and use more efficiently the sources they have, i.e. Japan has the lowest coal consumption per ton of iron in the world. Table 15 gives the energy requirement in different sectors in Japan.

TABLE 15

Estimates of Energy Requirements
According to the Main Sectors of Demand[1] in Japan

	Million tons of coal equivalent				
	1950	1960	1964	1970	1980
Final Consumption[3] in:					
Iron and steel	5.1	15.3	24.8	34	44
Other Industry	15.0	36.7	48.1	87[2]	198[2]
Transport (including bunkers)	8.8	19.0	33.1	59	126
Domestic and Miscellaneous	7.0	15.6	26.8	36	51
Energy sector[4]	4.8	7.6	11.9	14[2]	23[2]
Total Final Consumption	40.7	94.2	144.7	230	442
Transformation losses & non-energy products[5]	9.0	29.1	47.1	110	220
Total Primary Energy Requirements	49.7	123.3	191.8	340	662
of which for:					
Electricity Generation[6]	12.1	37.0	56.8	106	217
Specific Demand for:					
a) Coking Coal	3.0	12.5	19.8	28	38

(continued on next page)

Table 15 (continued)

	1950	1960	1964	1970	1980
b) Petroleum Products (for transport, bunkers & non-energy products)	1.5	15.2	35.3	73	168

1. The projections given do not represent any official view but are merely tentative estimates based on the extrapolition of recent trends.

2. Partly re-allocated, Secretariat estimates.
3. See Annex II, paragraphs 2-4. See Source for details.
4. i.e. not fuel for transformation but e.g. coal used at mines, oils used at refineries etc. and transmission losses of electricity and gas.

5. Chemical feedstocks make up a significant portion of non-energy uses of hydro-carbons.
6. Primary energy requirements for total electricity production, nuclear, hydro and thermal.

Source: OECD Report of Energy.

TABLE 16

Japan Energy Supply and Demand
Expressed in Terms of the Main Primary Sources[1]

	Million tons of coal equivalent				
	1950[2]	1960[2]	1964[2]	1970	1980
Total Requirements	49.7	123.3	191.8	340	662
Supplied from Indigenous Resources	46.4	70.5	72.3	84	109
Coal and Lignite	39.7	59.0	57.5	66	66
Crude Oil	0.4	0.6	0.8	1	1
Natural Gas	---	1.3	2.8	4	9
Hydro-Power	6.3	9.6	11.2	12	13
Nuclear Power	---	---	---	1	20
Net Imports	3.3	52.8	119.5	256	553
Oil	2.8	44.1	105.7	236	530
Solid fuels	0.5	8.7	13.8	20	23

1. This picture is based on unofficial, preliminary estimates available when this report was drafted.
2. Stock changes of coal with indigenous resources of oil with imports.

Source: OECD Report on Energy

By examining Table 16, we can follow the growth of energy consumption by source in Japan since 1950. The growth in coal has been low. Consumption has been growing about four and two tenths percent per annum. This is probably due to more efficient use and the high costs of importing coal.

On the other hand, the other sources of energy have been growing at a considerably high rate. Natural gas averaged thirteen per cent per annum, electric eleven and nine tenths per cent, and petroleum nineteen and two tenths per cent.

Petroleum would have averaged higher if it hadn't been for the cutbacks in recent years. From 1960 to 1969, it averaged twenty-two per cent and reached a high of thirty-nine and eight tenths per cent in 1960. Electricity too had a slightly higher average growth rate during the 60's boom of 12.47 per cent, reaching its high of 16.5 per cent in 1960. Following this trend, natural gas, (including gas for blast furnaces), which averaged 14.8 per cent during the period, peaked at 25.1 per cent in 1960. In fact, use actually decreased in year 1971, a -1.0 per cent.

Coupled with the tremendous industrial growth has been a need for more and more energy sources for Japan. These growth trends should continue in the near future, given that political problems do not hinder them too much.

We can see that growth in imports of both coal products and petroleum products has grown at average rates of 18.9 per cent and 19.7 per cent respectively. The growth of imported coal is considerably higher than the growth of consumption though, it being 4.2 per cent. This is due to the increasing per cent of consumption of coal from foreign sources, due to the lack of natural resources in Japan. We can not expect resources to improve, so we can expect this percentage to necessarily become larger as time goes on.

Petroleum imports are different because their growth rate, 19.7, is very close to the consumption growth of 19.2 per cent. The percentage of imports to consumption has increased only a small amount, i.e., 55.6 per cent in 1959 to 58.9 per cent (the highest) in 1973. This is due to the fact that Japan has always had to import most of its oil, and therefore increases in consumption have to have proportionate increases in imports. It is highly unlikely that this relationship will ever change.

Total imports have been growing at an average annual rate of 19.2 per cent. This is very close to the average rate of growth of the country as a whole, approximately twenty per cent per year. Import of energy has become a large portion of the total imports. It is now almost forty per cent, whereas it was approximately thirty per cent in the late fifties and early sixties. These are figures on dollar amounts and therefore may reflect the upward trend in prices in energy sources.

Yugoslavia

In the years since 1952, the composition of the important branches of Yugoslavian industry has undergone substantial change. Coal, food, tobacco, wood, textile, and leather and footware industries held the dominant position in the

Yugoslavian industrial picture. Since then, however, the chemical, electrical, petroleum, pulp and paper, metal-using and rubber industries, and electric power production came to the picture and are now the leading factors of rapid industrial growth. This change is reflective of the higher degree of industrial development that has been attained in Yugoslavia. The role that these industries play in Yugoslavia's economic growth has become similar to that of the developed countries.

World energy situation is of crucial importance to Yugoslavia. In the drive for an industrial economy, the available supply of energy resources will play a vital role.

The level of technical and technological progress can to a great extent be judged by the use of electric energy, since expansion of mechanization and automation necessarily requires increased consumption of electricity.

Yugoslavia is among those nations with the highest rate of growth of production and consumption of electricity. In 1970, 5.2 times as much electric energy was produced as in 1956, compared to a world increase of 2.5 times; i.e., the Yugoslavian increase was twice the world figure. There has also been a sizeable increase in energy consumption in Yugoslavia compared to other countries. Compared to 1962, the 1968 consumption of electricity increased for various countries as follows: Yugoslavia - 21%; Belgium - 17%; Bulgaria - 16%; the Netherlands - 13%; Italy - 10%; Great Britain - 5%; and the Federal Republic of Germany - 3%.

Tables 17 and 18 gives production and consumption of energy in Yugoslavia for 1968-71 (quantities in 000 metric tons):

TABLE 17

Energy Production

	Total	Coal & Lignite	Crude Petroleum	Natural Gas	Hydro & Nuclear Electricity
1968	19,304	13,783	3,272	0,778	1,471
1969	19,304	13,588	3,539	0,972	1,842
1970	21,417	14,532	3,741	1,301	1,843
1971	23,878	15,804	3,886	1,485	1,956

TABLE 18

Energy Consumption

	Total	Solid Fuels	Liquid Fuels	Natural Gas	Hydro & Nuclear Electricity
1968	25,213	15,475	7,452	0,778	1,508
1969	26,377	15,464	8,067	0,972	1,874
1970	29,337	16,430	9,737	1,301	1,869
1971	33,084	17,878	11,786	1,485	1,935

It can be seen from these figures that while production and consumption of natural gas and hydro and nuclear electricity are approximately equivalent, consumption of solid fuels outweighs coal production and the production of crude petroleum in Yugoslavia is by far insufficient when compared to consumption. As a result, a large amount of oil must be imported to meet the country's needs. The amount of oil imported is steadily increasing. In 1966, 2.2 million tons of crude oil was imported and by 1970 this figure was up to 4.5 million tons.

The oil and natural gas reserves in Yugoslavia have, to date, not been sufficiently explored. At present, the Pannonian Basin is the only location where crude oil and natural gas are being extracted. What is needed is a more systematic approach to exploration in the Pannonian Basin and greater prospecting and starting new production in the Adriatic Sea and the coastal belt.

Yugoslavia imports most of its oil from Iran and Iraq. Table 19 gives imports and exports of coal and petroleum for the period 1969-72:

TABLE 19

($ 000 U.S. dollars)

Exports	1969	1970	1971	1972
Coal	4,676	7,264	2,467	5,069
Crude oil	9,340	12,065	14,962	8,841
Imports	**1969**	**1970**	**1971**	**1972**
Coal	23,932	25,573	42,253	38,898
Crude oil	51,057	65,838	97,303	82,479

The increasing industrialization of Yugoslavia is reflected in the rising coal and oil import figures. Looking back at the energy consumption and production table, it is evident that energy demand is rising at a faster rate than supply. Therefore, increasing amounts of increasingly expensive fuels must be imported to fill the country's needs. The effect on Yugoslavia's balance of payments is obvious. Unless steps are taken to increase production of energy producing minerals within the country, the need to import oil will be a major contributor to the increasing balance of payments deficit.

A comparison of Yugoslavia's oil consumption, past and projected, can be seen from the table following page. Yugoslavia's oil needs are estimated to increase from approximately 34.7 million barrels per year in 1968 to over 124 million barrels per year in 1980. Its consumption is small, however, in comparison to the developed countries. In 1980 it will only need about one-tenth of W. Germany's requirements, one-thirty third of the U.S.S.R.'s needs and only one-sixtieth of the oil used in the United States.

In relation to other Western European states, its needs are more similar. Its consumption is relatively close to Czechoslovakia, E. Germany, Hungary and Poland. This is not surprising since the stage of economic development is similar for these neighboring states. Romania, on the other hand, seems to be further ahead given that oil consumption may be used as an indicator of economic progress.

The prospect of energy in most of the developing countries is quite bleak. They do not have any appreciable energy reserves but unless they embark on a vigorous development effort (for which energy supply is a precondition), they will face serious economic and political problems. The excessive rise in oil prices do not present any insurmountable problem to the developed countries since much of this are recovered by high cost of export of manufacturing goods. But only way of survival for the developing countries is to form cartel in other raw materials supply and develop atomic power. The energy situation in each country is unique. We do not have space to discuss it for some other countries. We only present some data in Tables 20-28 regarding oil consumption in the world and in different regions.

TABLE 20

Consumption per head of population in 1962 and 1964

	Tons of coal equivalent	
	1962	1964
OECD North America	8.6	9.0
OECD Europe	2.8	3.2
Japan	1.6	2.0
OECD area as a whole	4.4	4.9
USSR and Eastern Europe	3.0	
Communist Asia	0.6	
Middle East and Africa	0.4	
Caribbean and South America	0.7	
Oceania	3.0	
Far East[1]	0.1	
World	1.4	

(continued)

Table 20 (continued)

1. Not including Japan and Communist Asia

Source: OECD Energy Project

TABLE 21

World Oil Consumption

Consumption in terms of total crude usage, including all inland product consumption, international bunkering of ships and aircraft, refinery fuel and onshore and off-shore military requirements.

	(1,000 b/d) 1968	1970	1975	1980
W. Europe	10,315	11,810	16,110	21,910
Africa	796	918	1,293	1,930
Middle East	798	1,002	1,312	1,890
Asia-Pacific	5,037	6,347	10,584	15,210
U.S.-Canada	14,420	15,850	19,470	23,150
Latin America	2,470	2,786	3,684	4,900
Total Free World	33,836	38,713	52,453	68,990
Soviet Nations	5,704	6,637	9,617	14,160
Total World	39,540	45,350	62,070	83,150

TABLE 22

United States-Canada

	(1,000 b/d) 1968	1970	1975	1980
United States	13,080	14,400	17,700	21,000
Canada	1,340	1,450	1,770	2,150
Total	14,420	15,850	19,470	23,150

TABLE 23

Middle East

	(1,000 b/d) 1968	1970	1975	1980
Bahrain	41.4	51.5	65.0	95.0
Iran	257.0	325.0	435.0	640.0
Iraq	60.0	73.0	93.0	115.0
Israel	76.9	110.0	134.0	210.0
Jordan	9.6	10.5	13.0	16.0
Kuwait	98.0	122.0	164.0	240.0
Lebanon	30.8	37.5	50.0	75.0
Muscat Oman	6.7	8.3	12.0	17.0
Neutral Zone	9.3	11.6	14.5	18.0

(continued on next page)

Table 23 (continued)

	(1,000 b/d) 1968	1970	1975	1980
Saudi Arabia	150.0	183.0	245.0	350.0
Syria	35.0	42.0	51.0	65.0
Others	23.1	27.3	35.5	49.0
Total	797.8	1,001.7	1,312.0	1,890.0

TABLE 24

Asia-Pacific

	(1,000 b/d) 1968	1970	1975	1980
Australia	452	505	675	900
Brunei-Malaysia-Singapore	205	256	440	760
Burma	20	24	37	60
Ceylon	30	36	52	80
Hong Kong	57	63	85	105
India	329	445	670	950
Indonesia	130	155	225	330
Japan	2,976	3,900	7,000	9,950
Korea, South	104	175	350	540
New Zealand	70	76	95	110
Pakistan	87	97	125	160
Philippines	137	160	240	360
Taiwan	82	110	200	325
Thailand	140	130	165	290
Viet Nam, South	155	140	115	140
Others	63	75	110	150
Total	5,037	6,347	10,584	15,210

TABLE 25

Soviet Nations

	(1,000 b/d) 1968	1970	1975	1980
Albania	9	10	12	15
Bulgaria	85	95	140	220
China	295	312	350	410
Czechoslovaki	145	180	300	465
E. Germany	115	140	215	335
Hungary	110	130	210	330
Poland	145	185	280	440
Romania	270	335	510	845
U.S.S.R.	4,530	5,250	7,600	11,100
Total	5,704	6,637	9,617	14,160

TABLE 26

Europe

(1,000 b/d)

	1968	1970	1975	1980
Austria	145	190	275	410
Belgium & Lux.	440	510	660	860
Denmark	275	330	480	690
Finland	175	195	310	490
France	1,500	1,760	2,250	2,900
Greece	127	150	230	360
Ireland	60	90	125	180
Italy	1,413	1,700	2,550	3,900
Netherlands	590	640	770	950
Norway	135	155	215	310
Portugal	70	95	165	280
Spain	410	520	910	1,600
Sweden	460	540	760	1,070
Switzerland	210	250	360	530
Turkey	135	170	250	360
United Kingdom	1,825	1,950	2,380	2,900
W. Germany	2,150	2,320	3,050	3,550
Yugoslavia	95	120	200	340
Other	100	125	170	230
Total	10,315	11,810	16,110	21,910

TABLE 27

Africa

(1,000 b/d)

	1968	1970	1975	1980
Algeria	42.5	47.7	64.0	95.0
Angola	14.0	16.4	23.0	35.0
Congo, Kin.	13.0	15.0	21.0	31.0
Egypt	122.0	140.0	200.0	285.0
Ethiopia	9.2	10.7	15.0	23.0
Gabon	3.2	3.7	5.0	8.0
Guang	13.2	15.5	22.0	34.0
Kenya	26.2	30.5	43.0	66.0
Liberia	7.2	8.2	11.5	17.0
Libya	16.6	19.6	30.0	47.0
Malagasy	9.7	11.0	15.5	23.0
Morocco	33.3	38.0	53.5	82.0
Mozambique	9.3	10.8	15.0	24.0
Nigeria	28.4	32.0	45.0	69.0
Sierra Leona	5.7	6.5	9.0	13.0
Sudan	13.0	14.8	20.5	30.0
Tanzania	10.7	12.3	17.5	26.0
Tunisia	21.6	25.2	35.0	54.0
Union South Africa	222.0	259.0	365.0	535.0
Zambia	7.6	8.7	12.0	18.0
Others	168.0	192.0	270.0	415.0
Total	796.4	917.6	1,292.5	1,930.0

TABLE 28

Latin America

(1,000 b/d)

	1968	1970	1975	1980
Argentina	387.4	435.0	595.0	815.0
Bolivia	11.0	12.6	15.9	22.0
Brazil	440.0	520.0	675.0	880.0
Chile	84.5	92.0	125.0	170.0
Colombia	100.0	110.0	165.0	250.0
Costa Rica	6.8	7.7	8.9	12.5
Cuba	104.0	108.0	112.4	114.5
Ecuador	19.0	21.6	28.7	37.0
El Salvador	8.0	9.0	12.0	16.0
Guatemala	17.0	18.5	23.0	30.5
Honduras	8.9	9.4	11.5	14.0
Jamaica	27.2	33.3	43.0	60.0
Mexico	415.5	457.0	630.0	870.0
Netherlands Antilles	120.0	123.0	125.0	128.0
Nicaragua	7.7	8.3	11.0	15.0
Panama	50.0	58.5	77.5	104.0
Panama C. Z.	61.0	65.5	75.0	85.0
Paraguay	4.8	5.3	7.0	9.5
Peru	100.0	117.0	150.0	180.0
Puerto Rico & V.I.	126.5	151.0	213.0	290.0
Trinidad	74.0	83.0	115.0	160.0
Uruguay	37.8	41.9	47.5	62.0
Venezuela	215.0	247.0	345.0	480.0
Other	44.1	51.3	73.0	95.0
Total	2,470.2	2,785.9	3,684.4	4,900.0

Closely related to the problem of energy is the question of environment. Considerable amount of literature exists regarding social, economic, political and managerial aspects of environmental degradation. As search of more energy sources intensifies, natural landscape is going to be affected increasingly. As low quality coal with more sulphur content is used in place of oil, the air and water pollution will increase.

We can identify generally five kinds of air pollutants, namely:

1) Particulates
2) Sulphur oxides
3) Hydrocarbon
4) Carbon monoxide
5) Nitrogen oxide

The effects of pollution on the environment are easily observed in urban landscape, obnoxious odors in the air, smog, unhealthy rivers and lakes, and the rapid deterioration of paint, metal, and clothing in polluted environments. What is not readily observable are the effects of pollution on the general health.

The effects of various environmental pollutants on health have been investigated for a long time. In many cases a high degree of correlation has been demonstrated to exist between increased levels of pollution and a decrease in the general health of the populace. The appearance of articles in popular magazines about the effects of pollution on health reflects a growing concern for this problem among the public. The problem is not one restricted strictly to industrialized areas, but is world wide in scope.

A recent article in Time magazine pointed out some of the more important effects of pollution on health. There are the dangerous effects from exposure to industry specific pollutants such as P.V.C. plastic, lead smelting effluents, arsenic from copper smelting and others, referred to as occupational disease. The effects from the generally poor quality of an environment contributed to by these and other pollutants are called environmental disease. An example of evidence to support environment's effect is the higher incidence of lung cancer in the northeast United States which has poorer air quality because of greater industrialization than the rest of the country. Lately it has been discovered that the oil extraction process from shale and the gasification of coal produce deadly byproducts. These byproducts, polycyclic hydrocarbons, are highly carcinogenic, and their presence in the atmosphere will increase as we look towards supplying our energy needs in petroleum from these resources.

Considerable amount of work has been done in the economics of environment following closely the area of welfare economics. The general principles of public good and criteria of Pareto optimality, social welfare maximization, GNP criteria, economic efficiency and equality have been related to air and water pollution, solid waste management, etc. For operational purposes, cost-benefit analysis has been conducted and several schemes of polluter pay strategies have been developed. Environmental laws have been enacted in many countries based on these principles.

Several standard management science techniques like input-output, linear programming, dynamic programming, system analysis, etc. have been used to analyse the interrelationship of economic activity and environmental pollution. Recently energy has been treated as a special category. But not much has been done in the field of regional and interregional linkage of energy and environment. Since in most countries, people are concentrated in a few focal points, and resources are not uniformly distributed, the problems are regional in nature. This book containing some of the papers presented in the International Conference on Regional Science, Energy and Environment, held at Leuven, Belgium, May 1975, hopefully fills some gap in this direction.

The paper by Cumberland et al. concerns efficient management of energy resources at the regional level involving complex interdependencies and trade-offs between environmental quality, regional priorities and interregional competition, and national objectives.

In this paper, electrical energy requirements are estimated for the State of Maryland under three alternative sets of assumptions by five-year periods to 1985. The environmental impact associated with each of these scenarios is computed and the relationships between energy requirements and economic development strategies are estimated. Policy alternatives are estimated at the regional and at the national level under various assumptions concerning the rate and composition of economic development.

The paper by Breimer and Molle deals with three factors viz.: energy, environment and regional economics. It is limited largely to the situation in Western Europe, other areas are taken into account only as far as necessary for the correct interpretation of the Western European developments.

The paper is divided into two major parts, viz.: one dealing with the period from 1950 to 1970, which was characterized by cheap energy and relative neglect of environmental factors, and another dealing with the period since 1970, characterized by a growing awareness of the impact notably of industrial developments on the environment, and by the "energy crisis".

The energy-related industries considered are oil refineries, petrochemical industry, iron and steelmaking, etc. Certain developments with regard to deep-sea ports get analogous treatment. Attention is paid as well to the complexes these industries have often formed around certain nodes.

In the paper by Murray Brown et al. an attempt is made to determine whether the oil price rise is a balancing or disequilibrating factor in Italy's development. Using sectoral regional income originating with respect to the number of employees in the region as an index of growth, divergences and other types of behavior of this index are described over period, 1951-1968. A brief discussion of relevant international trade theorems reveals difficulties in applying them to the present problems with the information they had available. An alternative approach is set out which uses estimates of regional supply distribution equations by sector, estimates derived from a regional-national model of Italy. Evaluating these equations by using the price rises caused by the increase in crude oil prices yields the effect on regional income per worker of the change in fuel prices. This procedure holds constant all other variables. For various reasons, this method yields the short-run effects on regional income per worker of the rise in oil prices. It is found that the index of regional income per worker for all nineteen regions in Italy is adversely affected by the oil price rise. However, regions with higher incomes per worker are generally affected most. At least in the short run, this seems to indicate that the oil prices rise contributed to the balancing of regional development in Italy.

Pachuri devised a dynamic model representing the demand for and supply of electrical energy within a specific region in the S.E. part of the U.S. The demand for electrical energy is broken down into three major sectors - residential, industrial and commercial - and fifteen sub-sectors based on distinct types of use of electricity. The demand for each individual customer is related to prices of electricity, natural gas, fuel oil and coal (only in the industrial sector). The price effects adopted in this study are based on estimates of own-price and cross-price elasticities determined by other authors, who have done extensive research on this subject. The growth of customers in various sub-sectors are estimated by the author using linear regression. The variables affecting changes in the number of customers within each sub-sector are found to be total personal income for the region, income per capita and population.

Relationships determining the total population of the region are also estimated. Birth rates are estimated to be dependent significantly on regional income per capita. Death rates follow a steady time-related trend and migration is determined by the difference between regional per capita income and U.S. per capita income on the one hand and growth in industrial employment on the other.

The technique followed in estimating and validating this model is two-fold. Some relationships are estimated econometrically as mentioned above, and others are estimated by simulation, using past data in comparison with the "predictions" from the model for the same period. A series of simulations are then carried out to obtain the time path of variables up to 1990 using the systems dynamics approach. In order to drive the mode, estimates of future per capita income for the U.S. and total personal income for the region were extracted from a study published by the U.S. Department of Commerce. Regional population is determined endogenously in this model.

Using the model a number of policy alternatives were examined to assess the problems in supplying electricity to meet projected demand. It was concluded that in view of the financial crisis facing the electrical energy sector in the U.S. and the outlook for escalating prices of inputs, efforts must be made to curb the growth in demand. Pricing policies and other economic alternatives to promote this effort were identified and investigated, and the adoption of seasonal pricing and flattening out of rate schedules found useful in reducing total demand and seasonal peaking of loads.

Lesuis and Muller consider the short term effects of a reduction of the energy supplies, especially oil-products, for the Dutch economy. For this purpose the technique of linear programming was applied to an interregional input-output model. Apart from the usual input-output restrictions a set of energy-restrictions has been added. Labour market considerations appear either as restrictions or are treated in the objective function.

Various possibilities are considered as to the substitution of kinds of energy. Moreover the policy implications of different assumptions on import and export of energy products, as well as reductions in their final domestic use are analyzed.

In the paper by Moore and Zoltners, it is anticipated that the United States will be faced with shortages of various vital resources in the years ahead. The 1973 oil shortage is a case in point and shortages of metals, agricultural products and water are predicted. This paper examines the potential use of mathematical programming as a tool for scarcity management.

An economic model is presented which applies linear programming to determine the most efficient distribution of petroleum products among industries and regions. It maximizes employment (minimizes unemployment) while (1) insuring a balanced economy via input-output constraints, (2) maintaining at least the minimum production of basic goods and services necessary to prevent abnormally high inflation which could be caused by shortages, (3) maintaining an adequate inventory and a reasonable import-export position, (4) insuring that regional economies do not become severely depressed via regional output and employment constraints, and (5) making sure that the industry capacity in each region is not exceeded. The model utilizes an input-output framework at both the national and regional level and makes adjustments for regional differences with an employment multiplier. Although this combined input-output linear programming model is very useful, it requires considerable amount of data which may not be available. One way will be to construct a simulation model.

Foell, et al. present such a comprehensive energy model involving environment, population, land use, etc. in the form of an energy information system for policy guidance. This paper dealing with the situation in Wisconsin demonstrates the linkage between energy and other factors in a regional and interregional setting.

Donnelly and Parhizgari consider the regional impact on Maryland's economic structure due to the increased price of energy. The special connection between pollution abatement and energy has been brought up by Parvin and Grammas' formal analysis.

Heesterman's paper contains a general economic analysis of the state of the world economy. It is argued that the raw materials' crisis had economic causes, but was also triggered off partly by the currency crisis. At the same time, it also has an environmentally desirable aspect. The large-scale use of raw materials ought to be discouraged by their high price. This is so, for geologically scarce materials in order to preserve their supplies, but restraint of the use of scarce and non-scarce materials alike is desirable because of their impact on the environment.

The incentive effect of a change in relative prices in favour of resource-rents also has a regional aspect. This is because increased fuel costs make transport of products and/or materials over a large and geographically dispersed market-area uneconomic, thereby discouraging large-scale production and metropolitan concentration. Unfortunately, a sudden shift in the world's income-distribution in favor of resource-rents gives rise to serious circulation problems, which have far-reaching social and economic consequences. The main recommendations towards overcoming the present crisis and the same time safeguarding the environment require a much greater degree of international cooperation and planning, and a shift of taxation from income tax to materials-depletion taxes.

Labys offers a valuable survey of commodity modeling literature and suggestions for future research. For a real situation, Nijkamp and Mastenbroek concentrate on optimal decision rules for regional development alternatives with environmental impact studies and multiple-criteria decision problems. The use of optimal control theory is proposed as an expedient to deal with dynamic development problems. Given a formal specification of the economic-ecological decision framework at hand, a set of optimal decision rules is derived.

Next, the foregoing theoretical model is applied to a regional development program in the Northern part of the Netherlands (viz. the Dollard), where a friction between development efforts in the field of water resources and environmental preservation became apparent. The various evaluation problems are analysed within the aforementioned framework of a dynamic control model, in which a modified cost-benefit analysis plays a major role. Finally, an evaluation of the method employed and an outline of future research is presented.

Resek and Provenzano consider the dangers of automotive lead emissions. The United States Environmental Protection Agency has proposed a phased reduction schedule for lead in gasoline. This paper presents an economic dollar assessment of the risks and benefits associated directly with lead based on two alternative potential government policies. These alternatives are: first, no control on use of lead; and second, total elimination of lead. These dollar values are computed for the State of Illinois and for the United States.

The analysis computes first the cost of the removing of lead from gasoline (or benefits from continued use of lead). This requires technological analysis of changed costs given lead removal and ultimate effects of this on the supply schedule for gasoline. Second, it requires estimation of the gasoline demand schedule. This has been done in separate work by the authors and the results are presented in the paper. The lowest costs for Illinois under various assumptions are $112.8 million annually.

The risks of lead has been considered for three specific studies. First is reduction of agricultural yield. The second potential risk is the health hazards of lead contamination of agricultural products. Finally, the human health risks of urban dust contaminated by lead are considered. The total loss computed is very preliminary and speculative. For Illinois the estimated loss is $23 million annually. The benefits of lead in gasoline are found to greatly exceed the dollar value of risks, but future research is encouraged to make benefits and risks more specific.

Von Bremen's note discusses the link between energy use and some essential characteristics of technical progress in agriculture during the recent decades. Induced by the relative scarcity of factors in industrialized countries an intensive change of input structure has occurred. Secondly, a special kind of technical progress has evolved which can be characterized as labour-saving and energy-using. Thirdly, the technological change in the industrialized countries has implied rapidly diminishing possibilities of factor substitution, a fact which has been confirmed by econometric models estimating the elasticity of substitution between important groups of inputs.

Taking into account the points mentioned above he states that modern agricultural technologies cannot guarantee future food supplies if there is no security in the supply of energy, especially with hydrocarbons. Lack of energy endangers nearly all industrialized countries and those of the developing countries where the energy-consuming technologies are applied. Looking forward to a situation of energy shortages and high prices, especially with respect to hydrocarbons short-run and long-run possibilities of adjustment are discussed. Music outlines some impact on environmental design due to energy shortage.

Some of the above papers have analysed energy situations and recommended policies in respective countries. Since we live in an interdependent society, what we need is a world energy policy over and above national interests. What is clear is that the developed countries cannot keep on using and wasting energy in the scale as before, the developing countries cannot pay the exorbitant prices of energy, and there is a need of a mechanism to channel the accumulated revenue of the oil producing countries without disrupting the world monetary system.

To solve these problems, what we need is cooperation and not confrontation. Devising policies by the so-called developed and developing countries as separate groups might serve short-term objectives but it will not solve long-term problems. There is also the crucial need of developing new theories and methodologies to estimate the impact of shortages and the implications of a proposed policy. The traditional location theory may have to be modified, new concepts involving energy have to be added to input-output analysis, programming models have to be more realistic, comparative cost analysis should consider social and political factors, interregional industrial complex analysis should be developed and more important, the implication of the use of nuclear energy must be thoroughly investigated. We hope future studies will be forthcoming in this direction.

FORECASTING ALTERNATIVE REGIONAL ELECTRIC ENERGY REQUIREMENTS AND ENVIRONMENTAL IMPACTS FOR MARYLAND, 1970-1990[1]

John H. Cumberland
William Donnelly
Charles S. Gibson, Jr.
and Charles E. Olson

I INTRODUCTION

Although recognition of the relationships and conflicts between energy, environment, and economic growth objectives has become international in scope, many of the most difficult issues must be resolved at the regional level. Debate over energy policy has revealed new sets of conflicts between national interests and local interests. Moreover, the selection of specific sites for power plants determines the major impact of environmental damage and also determines who will bear the costs and who will receive the benefits from increased energy supplies.

Provision for resolving some of these issues has been provided in energy legislation passed in the State of Maryland. As the result of serious environmental threats to the Chesapeake Bay from the arbitrary siting by an electric utility of the State's first nuclear power plant, the Power Plant Siting Act of 1971 now requires advanced planning and public review of power plant siting plans, of their expected environmental impacts, and of demand estimates on which the siting plans are based.[2]

This paper reports on forecasts of electric energy requirements, on environmental emissions expected from this energy production and upon the translation of these forecasts into specific power plant siting decisions. A final section deals with some policy implications of the forecasts.

II ENERGY FORECASTS, ENVIRONMENTAL IMPACTS AND SITING DECISIONS

A. Energy Forecasts

In an effort to utilize the best features of each type of model available at the University of Maryland, several different methodologies were tested in making the electric energy demand projections. Maximum industrial detail was obtained by

---

[1] This paper summarizes an unpublished report on "Electric Energy Forecasts and Planning for Maryland, 1970-1990," by John H. Cumberland, William A. Donnelly, Charles S. Gibson, Jr., Charles E. Olson, and Frederick M. Peterson.

[2] _Laws of Maryland_, 1971, "Power Plant Siting," Chapter 31 (Senate Bill 540).

using input-output relationships from the Clopper Almon Inforum Model.[3] Detailed regional disaggregation was then obtained by applying these energy coefficients to the Curtis Harris Multi-Region, Multi-industry Forecasting Model.[4] Separate econometric studies were conducted for the residential, agricultural, and commercial sectors.[5]

Finally, although these approaches yielded generally consistent results, final results were presented in terms of three alternative energy scenarios, reflecting the critical role which state energy and development policy can play, especially through the effects of taxes, environmental regulation and economic development strategy.

The first, a high growth scenario, was based upon projection of historic growth rates. The second scenario assumes reduced growth because of price effects observed even before the oil embargo, and the third assumes lower growth rates because of a modest state energy conservation and management policy.

Results of these investigations are presented in the accompanying tables. Table 1 gives the industrial energy use coefficients derived from the Almon model and from census data. These coefficients were inserted into the Harris model in order to derive state and county forecasts of electricity requirements in the industrial sectors, as shown in Table 2. Although Maryland has an extensive electric energy grid, county estimates are needed in order to evaluate controversial additions proposed to this grid in view of the increasing awareness of the environmental and aesthetic impacts of transmission lines.[6]

Figures from these industrial use forecasts were combined with econometric estimates for non-industrial sectors and other data to generate the total electric energy forecasts under three different scenarios as summarized in Table 3.

The very high compound annual growth rate of 9.4 percent in Scenario I was derived from historical energy growth rates of the 1960's, and at the time of the

---

[3]Clopper Almon, Jr., Margaret B. Buckler, Lawrence M. Horwitz, Thomas C. Reimbold, 1985: Interindustry Forecasts of The American Economy, Lexington, D.C. Heath and Company, 1974.

[4]Curtis C. Harris, Jr., The Urban Economies, Lexington, D.C. Heath and Company, 1973.

[5]The authors are grateful to Dr. Frederick M. Peterson of the University of Maryland for these studies.

[6]Estimates were also made for each county of electricity requirements by industry, but are not reproduced here because of space limitations.

study represented the official estimate of the state public utility regulatory commission. The commission derived its estimates largely from totalling the industrial company forecasts and construction plans. These estimates, which were found to be excessively high, are now being modified by the companies and by the commission.

In Scenario II, the compound annual growth rates are estimated as 7.0 percent from 1970 to 1980 and 6.3 percent from 1980 to 1990. These reduced rates reflect the effects of price increases and other impacts which could be observed even before the energy crisis, and represent a realistic maximum level of electric energy requirements which can be expected over the period.

Scenario III forecasts a reduction of electric energy growth rates to 6.3 percent from 1970-1980, and 5.0 percent from 1980-1990. These rates are based upon the assumption of a minimal state plan for energy management and conservation, and are still higher than forecasts of the Federal Energy Administration.

B. Environmental Impacts

The next step in the analysis was to estimate environmental impacts associated with these alternative scenarios, for both nuclear and conventional plants. Environmental impacts were computed in terms of the emission of particulates, sulphur oxides, nitrogen oxides, carbon monoxide, and hydrocarbons into the air, thermal releases into the water, and radionuclides into air, water, and land.

Emission coefficients for airborne pollutants are shown in Table 4 for 1970, and combined for 1980 and 1990. Thermal emission coefficients are given in Table 5 for both nuclear and conventional plants for 1970, 1980, and 1990. Emission coefficients for radionuclides associated with nuclear power plants for 1970 and 1980 are presented in Tables 6 and 7.

These emission coefficients were then inserted into the model so that total emissions could be computed for each scenario. In order to translate these emission figures into environmental impacts it will be necessary to relate each emission at each plant site to the assimilative capacity of the local air, and water resources.[7]

The next step in the analysis was to summarize the energy requirements and accompanying environmental impacts for each of the three scenarios in Table 8.[8]

---

[7]See, for example, John H. Cumberland, "Economic Analysis in the Evaluation and Management of Estuaries," U.S. Environmental Protection Agency Report on Estuaries, forthcoming.

[8]Because of space limitations, estimates for release of radionuclides were limited to tritium only in Table 8.

C. Site Selection

The final purpose of the research was to translate the energy and environmental impacts of each scenario into actual decisions on the time sequencing of specific power plants. This step required relating energy requirements to peak load estimates and reserve ratios. Load factors and reserve ratios were used to translate energy requirements into capacity requirements for each scenario, for 1970, 1980, and 1990, as shown in Tables 9, 10, and 11. These estimates were then translated into annual forecasts of capacity required in Table 12, and into differences between planned and required additions to capacity for each year, for each scenario, in Table 13.

Finally, it was then possible to translate the above data into actual time schedules for construction of specific plants needed, in Table 14 for Scenario II and in Table 15 for Scenario III. The use of a formal siting model based upon local environmental assimulative capacities, air and water flows, aesthetic factors, and other environmental features would have made it possible to choose sites in order to minimize total environmental damage. In absence of such a siting model, the scheduling of plant construction for Scenario's II and III in Tables 14 and 15 was based on judgemental grounds. While these are not the only, or necessarily even the best schedules of plant construction, they do indicate that under the assumed conditions the original plans for at least two major nuclear power plants in Maryland could be postponed, and two eliminated for the foreseeable future. With a minimal state plan for energy conservation and management, even more reduction of costly plant construction and its associated environmental damage could be achieved.

III POLICY IMPLICATIONS

In the Power Plant Siting Act of 1971, Maryland has pioneered in the development of a State energy policy. But whether that policy turns out to be an innovative development to advance the public interest or merely a convenience to the utility companies for acquiring power plant sites will depend upon the administration of the Act, and upon the quality of the research which guides its administration.

The most important conclusion to be drawn from the limited data and trends examined in this study is that the era of uncritical acceleration of energy growth has ended in regional as in national and international economies. New factors which have emerged are now affecting both the demand for and the supply of electrical energy in the State. The most immediate effect of these changing forces is to reduce the urgency for initiating and constructing new electircal energy plants. Some installations, which until recently have been identified as essential, have now been postponed or cancelled by the public utilities themselves, and other energy facilities can be added to the list of those of which are certainly postponable and possibly unnecessary for the foreseeable future. Positive state action to reduce wasteful uses of energy and to begin emphasizing the quality of economic development, rather than the quantity of growth, can reduce the number of new electricity plants required even more. For example, by 1980 and 1990, the addition of each new increment of about 350,000 population in a region can trigger off the costly and controversial process of selecting a site for a new power plant. Planners of economic development and land use should be aware of this and other environmental implications of policies which add to population growth. Planning for economic growth, land use, and population size at the regional level can ease or aggravate energy problems.

Previous estimates of energy requirements made by the public utilities and largely accepted by the regulatory commissions have now been sharply reduced and can probably be scaled down even more in the future. Even more striking reductions could be effected without serious decreases in overall growth or welfare by commitment on the part of the region to exercise powers within its authority to insure that future economic growth strategy emphasizes the attraction of types of economic activity which generate high levels of economic welfare and low levels of energy use and environmental damage.

These are numerous steps that regions can take in order to establish a responsible energy policy. One would be to modify rate structures which now subsidize large consumers and to consider making rate structures progressive, rather than regressive, in order to discourage large consumers of energy and to reflect the rising marginal costs of energy. This would have the effect of avoiding the waste of energy and encouraging energy consumers to adopt efficient energy saving technologies. Another would be to adopt peak load pricing as an alternative to the present declining block type rate structures. Higher rates during peak periods would serve to discourage usage at such times and permit utilities to attain higher load factors. They would also delay significantly the need for new capacity that is occasioned by ever growing weather-sensitive peaks. The danger here is higher usage of existing plant and greater emission levels.[9]

Another important element in energy policy would be to impose taxes, environmental penalties, or emission charges, on the releases of conventional or nuclear pollutants into land, air, and water. Adoption of pollution penalties or emission charges could have many benefits. By making energy prices reflect the total real cost of pollution and production they encourage consumers to conserve,

---

[9] See Charles E. Olson, "Reforming Electricity Rate Structures in the United States," 93 Public Utilities Fortnightly, Volume 4 (February 14, 1974).

not waste energy. These charges also give energy producers financial incentives to design and install low-pollution technology. Emissions charges also have the advantage of generating public revenues which can be used for research on solving energy problems, compensating those who are damaged by power plants, or for general purposes.[10] Maryland, for example, already has enacted legislation which could lead to establishment of emission charges. The environmental surcharge on energy, now in effect, is a simple ad valorum tax. It now offers no incentives to reduce pollution. However, it could be converted to an emission charge or pollution tax by setting up a schedule of charges per unit of particulates, sulphur oxide, hydrocarbon, nitrous oxide, heat, decible of noise, and curie of radioactivity released.

The concept of an environmental emission charge could be broadened to reduce scenic and aesthetic damage, by imposing an annual penalty on visual pollution for every vertical foot of smokestack, cooling tower, and power plant and for every linear foot of transmission line per acre from which it was visible. Depending upon the rates charges, these measures could provide strong financial incentives for careful design and siting of power plants and transmission facilities.

Another important element in regional policy for energy conservation and management would be to adopt a regional economic development strategy which would encourage the adoption of land use plans, urban design, and transportation systems which would conserve energy and reverse the historical trends which have encouraged both the waste of energy and the design of transport systems which are highly dependent upon the lavish use of energy. One heritage which many large cities and communities now face is the extreme dependence on low-cost energy and great vulnerability to interruptions in energy supply. Rather than encourage accelerated development of industrial complexes and urban sprawl, the emergence of energy shortages should serve as a timely warning against continuing to attract more growth and more dependence upon energy supplies which in the future may be frequently interrupted. Thus, the "energy crisis" of the early 70's comes as a timely warning to adopt more responsible growth patterns in order to reduce vulnerability of individual homes, of communities, of transportation systems, and of urban concentrations to energy shortages and interruptions.

Responsible reaction to the energy crises can yield large future benefits to energy producers and consumers by avoiding the heavy costs of wasting energy, and by reducing the commitment of our land, air, and water resources to construction

---

[10] See, for example, John H. Cumberland, "Energy, Environment and Social Science Research Priorities," in Energy and Environment, Organization for Economic Cooperation and Development, Paris, 1974, pp. 1-50.

of energy facilities and transmission lines. Energy is probably the single greatest environmental problem both in the primary generation of energy, and in its subsequent use. For example, Japanese environmentalists are already expressing concern that even with extremely efficient and clean generation of energy, the consumption and use of energy and the heat inevitably generated by the use of energy may be approaching the critical level that in some areas could introduce large-scale climatological changes.[11] Release of heat, radioactivity and other emissions from production and consumption of energy are probably the ultimate environmental hazards from economic development. In addition to the thermal pollution problem, additional constraints on growth and energy consumption must ultimately result from the entropy principle, as emphasized by Georgescu-Roegen under which all forms of energy and material concentrations are being dissipated to non-recoverable levels.[12]

The petroleum crisis is generating widespread recognition that energy is the key variable and limiting factor in striking a reasonable balance between the desire for economic development and the maintenance of environmental responsibility towards the earth and its inhabitants. The total real costs of every new electric power plant are high, not only in terms of construction and operation, but also in terms of the environmental and aesthetic damage from the plant and transmission lines, not to mention the controversy, litigation, and psychic costs involved. Continued patterns of transportation, urban growth, and economic development based upon the erroneous assumption of cheap, uninterruptible energy supplies will leave society increasingly vulnerable to periods of energy shortage and interruption.

In summary, local governments have many actual and potential management options which they can use to tame the uncontrolled proliferation of electricity use and to help energy play a more constructive role in human welfare. Among the elements which could play an effective role in an improved state policy of energy conservation and management are elimination of rate structures which favor large energy-intensive consumers, heavy taxes on the emission of pollutants from power plants, taxes on aesthetic damage from power plants and transmission lines, and the adoption of an economic development policy to discourage activites which are wasteful consumers of energy and encourage activities which generate high levels of income and tax revenue per unit of energy used and pollution emitted.

The experience in Maryland also raises important issues of potential

---

[11] Yasuo Shimazu, "Energy Consumption and Limits to Growth in Japan," from Energy and Environment, Organization for Economic Cooperation and Development, Paris, 1974, pp. 239-271.

[12] Nicholas Georgescu-Roegen, "Energy and Economic Myths," Southern Economic Journal, Volume 6, Number 3, January, 1975, pp. 347-381.

conflict between local interests and interests at the state and federal levels. The Maryland Power Plant Siting Act and proposed federal legislation could be used to coerce local areas into accepting construction of large environmentally damaging energy facilities in local areas against local wishes, and at great local cost, on the basis of an assumed overriding higher interest. This is a very questionable approach for several reasons. First of all, it overlooks economic benefits which can result from the assertion of local rights. Permitting local vetoes of unwelcome energy prospects can provide strong incentives to develop less damaging energy technologies. Permitting local vetoes of damaging energy facilities could also leave open the option of permitting external beneficiaries of the energy prospects to pay compensation to the damaged region. This solution has attractions both from the point of view of equity and efficienty. Secondly, central coercion of local areas is politically and ethically questionable in a federal system.

The risk of occasional brownouts and shortages can be met by application of well-designed peak-load pricing structures and prohibition of new large hook-ups which do not clearly serve the public interest. The legitimate demands for energy for continuing economic development consistent with environmental responsibility can most effectively be met by adopting a carefully planned set of pollution penalties, emission taxes, and aesthetic damage charges plus an economic development policy setting priorities for the types of economic expansion which increase economic welfare without excessive environmental damage. Emission charges, energy surtaxes, and limited prohibition of some types of large new hook-ups would affect some development interests primarily by discouraging large-scale energy waste but would have the advantage of encouraging a high quality of economic development. Cancellation or postponement of additional electric plant construction as recommended in this study will also significantly reduce pollution emissions.

Taking these modest steps towards development of a policy on energy management and economic development could make it possible to eliminate entirely or to postpone indefinitely (until possibly better technologies are available) the construction of some of the costly and damaging plants now being considered, not only without serious sacrifices in energy supply, but with probably increases in economic welfare and improvements in the quality of life.

TABLE 1

Industrial Electrical Use Coefficients
(Kilowatt Hours Per Dollar of Value Added)
(1970 Prices)

Industry	Coefficient	Industry	Coefficient
05 Iron Ore Mining	9.5540	31 Office Furniture	0.54270
06 Non-Ferrous Ore Mining	4.6400	32 Paper and Proc., Excl. Containers	5.3390
07 Coal Mining	3.6100	33 Paper Containers	0.64870
08 Petroleum Mining	2.2730	34 Printing and Publishing	0.40090
09 Minerals Mining	4.2310	35 Basic Chemicals	10.016
10 Chemical Mining	12.718	36 Plastics and Synthetics	3.1540
11 New Construction	0.0000	37 Drugs, Cleaning and Toilet Items	0.46530
12 Maintenance Constr.	0.0000	38 Paint and Allied Products	0.32860
13 Ordinance	0.79500	39 Petroleum Refining	1.2370
14 Meat Packing	0.18080	40 Rubber and Plastic Products	1.7080
15. Dairy Products	0.41970	41 Leather Tanning	0.95790
16 Canned and Frozen Foods	0.45070	42 Shoes and Other Leather Products	0.30780
17 Grain Mill Products	0.57410	43 Glass and Glass Products	3.0910
18 Bakery Products	0.49350	44 Stone and Clay Products	2.8440
19 Sugar	0.73210	45 Iron and Steel	3.9440
20 Candy	0.50590	46 Copper	0.80440
21 Beverages	0.40130	47 Aluminum	16.390
22 Misc. Food Products	0.52520	48 Other Non-Ferrous Metals	2.0310
23 Tobacco	0.18520	49 Metal Containers	0.53750
24 Fabrics and Yarn	1.9330	50 Heating, Plumbing, and Struc. Metal	0.60850
25 Rugs, Tire Cord and Misc. Textiles	0.64710	51 Stampings, Screw Mach. Prod.	0.97700
26 Apparel	0.28650		
27 Household Textiles and Upholst.	0.26980		
28 Lumber and Prod., Excl. Containers	0.87490		
29 Wooden Containers	0.46440		
30 Household Furniture	0.56180		

TABLE 2

Maryland Industrial Electrical Demand Forecasts by County for Selected Years from 1970 through 1990 (Megawatt Hours)

County	1970	1975	1980	1985	1990
Allegany	338,143	517,315	738,136	975,309	1,287,770
Anne Arundel	156,675	209,830	289,735	376,493	484,721
Baltimore	1,173,880	1,708,130	2,245,290	2,754,520	3,396,180
Baltimore City	1,550,800	2,300,880	3,254,160	4,307,670	5,716,440
Calvert	1,756	2,664	3,768	4,925	6,302
Caroline	7,976	11,649	16,647	22,702	31,179
Carroll	62,707	95,056	122,574	146,199	171,508
Cecil	92,851	105,064	118,538	130,216	148,133
Charles	10,442	14,648	19,907	25,216	31,592
Dorchester	17,118	29,981	46,233	61,809	79,871
Frederick	110,119	204,734	374,139	653,056	1,024,160
Garrett	8,542	8,880	9,508	10,368	11,780
Harford	145,363	244,086	370,255	501,089	655,646
Howard	66,503	84,588	105,366	127,110	159,016
Kent	10,306	14,486	20,058	25,421	31,811
Montgomery	79,961	104,535	148,401	204,344	267,512
Prince George's	144,765	179,013	220,853	267,443	332,886
Queen Anne's	2,783	3,385	4,024	4,579	5,336
St. Mary's	1,756	2,120	2,409	2,579	2,874
Somerset	4,109	5,301	7,303	10,824	16,956
Talbot	9,153	13,815	20,208	27,146	34,676
Washington	105,165	127,781	151,647	171,238	195,202
Wicomico	23,109	32,772	46,171	60,758	79,093
Worcester	10,635	15,064	22,044	31,462	43,837
Maryland Total	4,134,617	6,035,776	8,357,374	10,902,476	14,214,487

Table 1 (continued)

Industry	Coefficient	Industry	Coefficient
52 Hardware, Plating, and Wire Prod.	0.91240	74 Misc. Manufactured Products	0.47530
53 Engines and Turbines	0.69480		
54 Farm Machinery and Equipment	0.48490		
55 Construction and Mining Mach.	0.77330		
56 Material Handling Equipment	0.41350		
57 Metalworking Mach. and Equipment	0.85380		
58 Special Industrial Machinery	0.55440		
59 General Industrial Machinery	0.74950		
60 Machine Shops and Misc. Machinery	1.0300		
61 Office and Computing Machines	0.47360		
62 Service Industry Mach.	0.38260		
63 Electric Apparatus and Motors	1.2510		
64 Household Appliances	0.69730		
65 Electric Light and Wiring Equipment	0.73100		
66 Communication Equip.	0.62690		
67 Electronic Components	1.1680		
68 Batteries and Engine Electrical Equipment	0.87710		
69 Motor Vehicles	0.42570		
70 Aircraft and Parts	0.70490		
71 Ships, Trains, Trailers and Cycles	0.47510		
72 Instruments and Clocks	0.63080		
73 Optical and Photo. Equipment	1.0760		

TABLE 3

A Comparison of the Forecasts of Maryland's
Electric Energy Requirements for 1980 and 1990
(Energy in Billions of KWH)

	Electric Energy Requirements			Compound Growth Rates	
Forecast	1970	1980	1990	1970-80	1980-90
Scenario I	23.2	57.1	140.0	9.4%	9.4%
Scenario II	23.3	45.7	84.0	7.0%	6.3%
Scenario III	23.3	43.0	70.0	6.3%	5.0%

TABLE 4

Steam Electric Plant - Emissions Coefficients

1970[1]

Pollutant	Emission Coefficients	
Particulates*	.030635	LBS/KWH
Sulfur oxides	.029384	LBS/KWH
Nitrogen oxides	.007315	LBS/KWH
Carbon Monoxide	.000141	LBS/KWH
Hydrocarbons	.000078	LBS/KWH

1980[1,2]

Pollutant	Emission Coefficients	
Particulates*	.028478	LBS/KWH
Sulfer oxides	.015333	LBS/KWH
Nitrogen oxides	.005529	LBS/KWH
Carbon monoxide	.000106	LBS/KWH
Hydrocarbons	.000059	LBS/KWH

1990[1,2]

Pollutant	Emission Coefficients	
Particulates*	.028478	LBS/KWH
Sulfer oxides	.015333	LBS/KWH
Nitrogen oxides	.005529	LBS/KWH
Carbon monoxide	.000106	LBS/KWH
Hydrocarbons	.000059	LBS/KWH

[1]These coefficients were derived from data supplied by the U.S. Environmental Protection Agency.

[2]These emissions assume the use of low-sulfer fuels and boiler operations such that NO_x, CO and HC are all reduced by 20 percent.

*These are gross emission coefficients. No attempt was made to estimate the net emission coefficients due to existing traetment facilities.

TABLE 5

Steam Electric Plant - Thermal Emission Coefficients[1]

Maryland

Conventional Year	Emission Coefficient
1970	7,493 BTU/KWH
1980	7,111 BTU/KWH
1990	6,743 BTU/KWH

Nuclear Year	Emission Coefficient
1970	7,564 BTU/KWH
1980	7,180 BTU/KWH
1990	6,915 BTU/KWH

[1]The emission coefficients were calculated by subtracting 3,413 BTU, the energy equivalent of one kilowatt-hour, from Maryland average heat rates given in Table 4.

TABLE 6

Nuclear Plant - Emission Coefficients

United States Emission Characteristics For
Nuclear Powered Electric Energy Plants

1970 Radioactive Isotope	Emission Coefficient*		
	Air Curies/KWH	Land Curies/KWH	Water Curies/KWH
H3	2.4277 -09	6.8849 - 09	8.0875 - 07
N13	1.3812 - 07	0.0000	2.8087 - 16
C14	9.5052 - 20	3.5062 - 20	2.9966 - 18
NA22	0.0000	2.2113 - 18	6.2870 - 20
NA24	0.0000	1.8566 - 38	3.4489 - 11
AR39	2.0305 - 13	0.0000	6.9257 - 17
AR41	2.2295 - 07	0.0000	4.9517 - 13
CR51	0.0000	6.3197 - 13	8.2375 - 12
MN54	0.0000	1.2931 - 09	1.2610 - 12
FE55	0.0000	4.6811 - 08	3.6604 - 11
FE59	0.0000	1.4301 - 11	1.9774 - 12
CO58	0.0000	3.1771 - 09	1.0406 - 10
CO60	0.0000	2.2469 - 08	8.0052 - 11
N163	0.0000	5.9792 - 09	1.8964 - 11
CU64	0.0000	9.0193 - 45	1.0908 - 18
ZN65	0.0000	1.6161 - 12	5.7993 - 15
KR85M	7.9635 - 06	0.0000	0.0000
KR85	6.7862 - 09	0.0000	0.0000
KR87	2.3851 - 05	0.0000	0.0000
KR88	2.6172 - 05	0.0000	0.0000
SR89	9.2502 - 15	2.8015 - 10	1.6258 - 11
SR90	1.0410 - 17	1.1664 - 10	9.8024 - 14
ZR95	2.7752 - 16	8.2783 - 11	2.1666 - 12
NB95	4.1363 - 16	1.7300 - 10	1.5073 - 13
MO99	1.0196 - 11	5.0252 - 36	1.9016 - 09

Table 6 (continued)

United States Emission Characteristics for Nuclear Powered Electric Energy Plants

1970 Radioactive Isotope	Emission Coefficients*		
	Air Curies/KWH	Land Curies/KWH	Water Curies/KWH
RU103	1.4853 - 12	3.2861 - 12	1.4184 - 11
RU106	1.3726 - 13	7.7289 - 08	1.3079 - 10
TE132	6.5365 - 12	3.2874 - 36	3.4106 - 10
I129	4.2582 - 19	1.2570 - 14	2.4675 - 18
I131	2.0048 - 11	8.6250 - 20	5.3458 - 12
I132	3.4583 - 11	8.3259 - 36	3.1611 - 10
I133	1.9795 - 12	6.6659 - 36	1.6344 - 10
I135	3.1417 - 11	8.3104 - 36	6.0711 - 11
XE131	1.5970 - 08	5.9176 - 17	0.0000
XE133M	2.9241 - 07	1.5620 - 20	0.0000
XE133	8.2110 - 06	1.1778 - 18	0.0000
XE135M	1.3284 - 05	0.0000	0.0000
XE135	5.1714 - 06	0.0000	0.0000
XE138	4.6365 - 07	0.0000	0.0000
CS134	4.6669 - 18	1.0976 - 08	9.2541 - 11
CS137	1.0824 - 14	4.0048 - 08	4.1624 - 11
BA140	4.0139 - 15	1.5376 - 16	1.6048 - 12
LA140	7.5100 - 17	1.7689 - 16	3.3091 - 12
CE141	8.7913 - 16	5.4808 - 12	2.0022 - 11
CE144	5.8165 - 16	3.6445 - 10	1.4286 - 12

*These coefficients were derived from data supplied to the Bureau of Business and Economic Research by the U.S. Atomic Energy Commission. As reported in Battelle Northwest Laboratories, data for preliminary demonstration of "Environmental Quality Information and Planning System (EQUIPS)" December, 1971.

TABLE 7

Nuclear Plant - Emissions Coefficients

United States Emission Characteristics for
Nuclear Powered Electric Energy Plants

1980 Radioactive Isotope	Emission Coefficient*		
	Air Curie/KWH	Land Curie/KWH	Water Curie/KWH
H3	2.8355 - 09	8.0426 - 09	9.3760 - 07
N13	8.1480 - 08	0.0000	3.2807 - 16
C14	7.6966 - 18	6.4129 - 20	1.2901 - 20
NA22	0.0000	2.3298 - 18	4.2598 - 21
NA24	0.0000	1.7440 - 38	4.6428 - 12
AR39	1.7594 - 13	0.0000	8.0894 - 17
AR41	1.3152 - 07	0.0000	5.7838 - 13
CR51	0.0000	6.6070 - 13	8.0701 - 13
MN54	0.0000	1.4829 - 09	6.7727 - 13
FE55	0.0000	5.2964 - 08	1.3879 - 11
FE59	0.0000	1.6315 - 11	8.8235 - 13
CO58	0.0000	3.5019 - 09	2.1582 - 11
CO60	0.0000	2.0530 - 08	9.3076 - 13
N163	0.0000	5.4319 - 09	2.0937 - 13
CU64	0.0000	8.6155 - 45	8.7732 - 19
ZN65	0.0000	1.7019 - 12	7.8201 - 16
KR85M	1.7724 - 06	0.0000	0.0000
KR85	6.6044 - 09	0.0000	0.0000
KR87	3.5228 - 06	0.0000	0.0000
KR88	4.3696 - 06	0.0000	0.0000
SR89	1.0199 - 14	3.5859 - 10	1.4958 - 12
SR90	1.0439 - 17	1.491 - 10	8.8532 - 15
ZR95	2.1455 - 16	9.6321 - 11	1.8696 - 13
NB95	4.3792 - 19	2.0129 - 10	7.7288 - 15
MO99	9.5459 - 12	6.5068 - 36	2.5794 - 10
RU103	1.4157 - 12	4.1222 - 09	1.2959 - 10

Table 7 (continued)

United States Emission Characteristics for Nuclear Powered Electric Energy Plants

1980 Radioactive Isotopes	Emission Coefficients*		
	Air Curie/KWH	Land Curie/KWH	Water Curie/KWH
RU106	1.2631 - 13	9.4034 - 08	1.0877 - 11
TE132	2.1215 - 12	4.2453 - 36	1.8175 - 10
I129	4.9520 - 18	1.4790 - 14	7.0764 - 19
I131	2.1756 - 11	1.1043 - 19	1.9603 - 10
I132	6.2930 - 12	1.1737 - 35	1.4603 - 10
I133	2.9787 - 12	9.0343 - 36	1.4530 - 10
I135	4.6339 - 12	1.1659 - 35	7.0836 - 11
XE131	1.1975 - 08	6.9474 - 17	0.0000
XE133M	7.0885 - 08	2.1165 - 20	0.0000
XE133	3.9854 - 06	1.5962 - 18	0.0000
XE135M	1.9474 - 06	0.0000	0.0000
XE135	7.5916 - 07	0.0000	0.0000
XE138	6.8380 - 06	0.0000	0.0000
CS134	3.5712 - 15	1.2668 - 08	2.7781 - 12
CS137	8.9641 - 15	4.7746 - 08	1.7956 - 12
BA140	3.8131 - 15	1.9756 - 16	1.5655 - 13
LA140	8.2000 - 17	2.2730 - 16	3.8021 - 13
CE141	8.2566 - 16	7.0227 - 12	1.8654 - 12
CE144	5.5189 - 17	4.6617 - 10	1.2965 - 13

*These coefficients were derived from data supplied by the U.S. Atomic Energy Commission. As reported in Battelle Northwest Laboratories, data for preliminary demonstration of "Environmental Quality Information and Planning System (EQUIPS)" December, 1971.

TABLE 8

A Comparison of Pollution Emissions Generated Directly By Alternative Energy Production Levels In Maryland

	Scenario I High Growth Trend Projection	Scenario II With Effects of Price Rises	Scenario III With State Energy Conservation and Management Policies
1970			
Energy (Billion KWH)	23.3	23.3	23.3
Particulates (tons)	3.57 + .05	3.57 + .05	3.57 + .05
Sulpher oxides (tons)	3.42 + .05	3.42 + .05	3.42 + .05
Nitrogen oxide (tons)	8.52 + .04	8.52 + .04	8.52 + .04
Carbon monoxide (tons)	1.64 + .03	1.64 + .03	1.64 + .03
Hydrocarbons (tons)	9.09 + .02	9.09 + .02	9.09 + .02
Heat (BTU's)	17.46 + .13	17.46 + .13	17.46 + .13
Radioactive Tritium (curies)	0.00	0.00	0.00
1980			
Energy (Billion KWH)	57.1	45.7	43.0
Particulates (tons)	5.05 + .05	4.67 + .05	4.41 + .05
Sulphur oxides (tons)	3.15 + .05	2.51 + .05	2.38 + .05
Nitrogen oxide (tons)	1.14 + .05	9.07 + .04	8.57 + .04
Carbon monoxide (tons)	2.18 + .03	1.74 + .03	1.64 + .03
Hydrocarbons (tons)	1.21 + .03	9.68 + .02	9.15 + .02
Heat (BTU's)	41.33 + .13	33.0 + .13	31.12 + .13
Radioactive Tritium (curies)	15.17 + .03	12.14 + .03	11.38 + .03
1990			
Energy (Billion KWH)	140.0	84.0	70.0
Particulates (tons)	9.97 + .05	5.98 + .05	4.98 + .05
Sulphur oxides (tons)	5.37 + .05	3.22 + .05	2.68 + .05
Nitrogen oxide (tons)	1.94 + .05	1.16 + .05	9.68 + .04
Carbon monoxide (tons)	3.71 + .03	2.23 + .03	1.86 + .03
Hydrocarbons (tons)	2.07 + .03	1.24 + .03	1.03 + .03
Heat (BTU's)	95.61 + .13	57.36 + .13	47.8 + .03
Radioactive Tritium(curies)	66.39 + .03	39.84 + .03	33.11 + .03

TABLE 9

Scenario I Forecasts of Generating Capacity Required in Maryland for Selected Years

	1970
Energy Demand Forecast	23.3 Billion KWH
Load Factor	52.4%
Peak Load Forecast	5054MW
Reserve Ratio	21%
Generating Capacity Required	6166 MW
	1980
Energy Demand Forecast	57.1 Billion KWH
Load Factor	52.7%
Peak Load Forecast	12335MW
Reserve Ratio	21%
Generating Capacity Required	14925MW
	1990
Energy Demand Forecast	140 Billion KWH
Load Factor	53%
Peak Load Forecast	30154MW
Reserve Ratio	21%
Generating Capacity Required	36487MW

TABLE 10

Scenario II Forecasts of Generating Capacity Required in Maryland for Selected Years

	1970
Energy Demand Forecast	23.3 Billion KWH
Load Factor	52.4%
Peak Load Forecast	5054MW
Reserve Ratio	21%
Generating Capacity Required	6116MW
	1980
Energy Demand Forecast	45.7 Billion KWH
Load Factor	52.7%
Peak Load Forecast	9872MW
Reserve Ratio	21%
Generating Capacity Required	11945MW
	1990
Energy Demand Forecast	84 Billion KWH
Load Factor	53%
Peak Load Forecast	18093MW
Reserve Ratio	21%
Generating Capacity Required	21892MW

TABLE 11

Scenario III Forecasts of Generating Capacity Required in Maryland for Selected Years

	1970
Energy Demand Forecast	23.3 Billion KWH
Load Factor	52.4%
Peak Load Forecast	5054MW
Reserve Ratio	21%
Generating Capacity Required	6116MW
	1980
Energy Demand Forecast	43.0 Billion KWH
Load Factor	52.7%
Peak Load Forecast	9289MW
Reserve Ratio	21%
Generating Capacity Required	11240MW
	1990
Energy Demand Forecast	70.0 Billion KWH
Load Factor	53%
Peak Load Forecast	15077MW
Reserve Ratio	21%
Generating Capacity Required	18243MW

TABLE 12

Generating Capacity Required 1970-1984

Year	Annual Generating Capacity Required		
	Scenario I*	Scenario II**	Scenario III***
1970	6,116MW	6,116MW	6,116MW
1971	6,687MW	6,539MW	6,500MW
1972	7,311MW	6,992MW	6,908MW
1973	7,993MW	7,476MW	7,341MW
1974	8,739MW	7,993MW	7,802MW
1975	9,554MW	8,547MW	8,291MW
1976	10,446MW	9,139MW	8,811MW
1977	11,420MW	9,772MW	9,364MW
1978	12,486MW	10,448MW	9,952MW
1979	13,651MW	11,172MW	10,576MW
1980	14,925MW	11,945MW	11,240MW
1981	16,317MW	12,772MW	11,945MW
1982	17,840MW	13,656MW	12,695MW
1983	19,505MW	14,602MW	13,491MW
1984	21,325MW	15,612MW	14,337MW

* Based upon a 9.3% compound annual growth rate
** Based upon a 6.92% compound annual growth rate
*** Based upon a 6.27% compound annual growth rate

TABLE 13

Differences Between Planned Additions to Capacity
and Scenario I, II and III's Required Additions to Capacity

Year During Which There Would Be A Capacity Shortage or Surplus	Scenario I*	Scenario II*	Scenario III*
Jan. 1, 1975	+ 645	+ 906	+ 971
Jan. 1, 1976	+ 553	+1114	+1251
Jan. 1, 1977	+ 259	+1161	+1378
Jan. 1, 1978	- 207	+1085	+1390
Jan. 1, 1979	- 772	+ 961	+1366
Jan. 1, 1980	- 946	+1288	+1802
Jan. 1, 1981	-2338	+ 468	+1097
Jan. 1, 1982	-2761	+ 677	+1447
Jan. 1, 1983	-2166	+1991	+2911
Jan. 1, 1984	- 326	+4641	+5725

*A positive sign indicates that if the presently proposed plant schedule proceeded there would be excess generating capacity. A negative sign indicates a shortage of generating capacity.

TABLE 14

A Recommended Plant Schedule
Based Upon Scenario II's Demand Forecast
(1974-1983)

Plant	Unit	Capacity	Year to be Placed in Service
Chalk Point	#3	660MW	1975
Calvert Cliffs	#1	800MW	1975
Chalk Point	#4	660MW	1977
Calvert Cliffs	#2	800MW	1977
Brandon Shores	#1	600MW	1978
Brandon Shores	#2	600MW	1979
Douglas Point	#1	1100MW	1980
Douglas Point	#2	1100MW	1981
Dickerson	#4	800MW	1982
Unknown		600MW	1983
Postpone:	Calvert Cliffs	#1	1 year
	Calvert Cliffs	#2	2 years
	Douglas Point	#1	1 year
Cancel:	Canal	#1	
	Canal	#2	
	Point of Rocks		
Add:	600MW Units		1983

TABLE 15

A Recommended Plant Schedule
Based Upon Scenario III's Demand Forecast
(1974-1983)

Plant	Unit	Capacity	Year to Be Placed in Service
Chalk Point	#3	660MW	1974
Calvert Cliffs	#1	800MW	1975
Chalk Point	#4	660MW	1976
Calvert Cliffs	#2	800MW	1977
Brandon Shores	#1	600MW	1979
Brandon Shores	#2	600MW	1980
Unknown		25MW	1980
Station J		1000MW	1981
Canal	#1	1160MW	1982
Canal	#2	1160MW	1983
Postpone:	Calvert Cliffs	#1	1 year
	Calvert Cliffs	#2	2 years
	Brandon Shores	#1	2 years
	Brandon Shores	#2	2 years
Cancel:	Douglas Point	#1	
	Douglas Point	#2	
	Point of Rocks		
Add:	25 MW Units		1980

THE LOCATION OF ENERGY-RELATED INDUSTRIES IN WESTERN EUROPE

W. T. Molle and Henk Breimer

I INTRODUCTION

The location of energy-related industries has been getting a lot of attention since the recent "energy-crisis", particularly in the countries of Western Europe, which have to rely heavily on supplies of energy from outside their area. Considerations of military strategy, balance of payments, and regional development were among the reasons for the increased interest. In the present article the historical location pattern of some selected energy-related industries in Western Europe will be analysed with a view to discovering the changes wrought by outside factors on-location patterns. For practical reasons the study has been confined to the period after the Second World War.

A first section will deal with the development of the energy market, distinguishing sources of energy and sectors of consumption. Next, it will be reported how oil refineries, petrochemical industry and iron and steel works have been selected, by practical criteria, as the most important energy-related industries. The location pattern of these industries will be analysed in separate sections, in which, after a short theoretical and historical description of the industry's location pattern, an attempt will be made to test the patterns theoretically derived, with the help of statistical material for the period 1950-1970 and, occasionally, a simple model. Oil refineries as well as petrochemical industries were found to cluster, and to locate in the market for their products. Iron and steel works were also found to orientate to the market. The only orientation towards energy deposits that could be established was a tendency to locate near the sea coast, but this tendency affects only the choice of sites within a national market.

II SELECTION OF ENERGY-RELATED INDUSTRIES

The pattern of energy consumption and production in Western Europe has shown dramatic changes in the last few decades. The high increase in consumption is particularly striking. From only one per cent per annum between 1930 and 1950, it rose to 3.8 per cent per annum between 1950 and 1960 (OECD 1966), to attain a level of 5.5 per cent per annum in the period 1960-1970 (OECD 1973). This overall growth was distributed unequally among the sources of primary energy. A good indication of the relative positions of those sources and the shifts in these positions over time is given by the shares in total energy consumption.

Table 1. Consumption of energy in OECD Europe expressed in terms of the main primary sources (percentages)

Year	Solid fuels	Natural gas	Oil	Nuclear energy	Hydro power	Total
1929	95	-	4	-	1	100
1937	90	-	8	-	2	100
1950	83	-	14	-	3	100
1955	74	1	22	-	3	100
1960	61	2	33	-	4	100
1965	45	2	48	1	4	100
1970	29	7	60	1	3	100

Source: OECD 1961 and 1973

One observes how coal, from being almost the sole source of energy around 1930, has been steadily reduced to a modest share of 30 per cent in 1970, whereas oil consumption has risen very steeply, to arrive at a share of 60 per cent in 1970. Compared to coal and oil, hydro power and nuclear energy occupy very minor positions, as does gas, although it is on the up and up.

To the changes in position of the primary sources of energy in relation to total energy consumption, the various energy users have contributed to different degrees. In the OECD Energy Statistics - the only source available that gives internationally comparable figures for the area of study - three sectors are distinguished as major energy consumers, viz. 'transport', 'industry', and 'other', the last-mentioned sector comprising mainly the domestic and commercial users. All the figures from these statistics have been converted to oil equivalents to make them comparable, by means of conversion factors as given in OECD (1973). As sources of energy have been distinguished only solid fuels (mainly coal), gaseous fuels (manufactured and natural gas), liquid fuels (mainly oil), and electricity. Electricity can be generated from any of the primary sources, and there are no statistics available about the shares of each of them. It is, therefore, impossible to translate electricity figures into primary-source figures; consequently, they are given as such. The consumption of energy by the energy sector itself, which is quite large (comprising power for production and transmission as well as losses), is not included in the figures to be presented here.

In the OECD energy statistics, separate figures are given for the final energy consumption by the iron and steel industry; from the yearly OECD publications on the chemical industry comparable figures of the final energy consumption by the chemical industry could be drawn. However, these data were available only for the ten largest production countries in Western Europe (accounting for approximately 95 per cent (OECD-e) of the total turnover of the Western European chemical indsutry), and for the years since 1961.

To trace some historical developments, figures for the benchmark years of five-year periods between 1950 and 1970 have been worked out and analysed. It appeared that for all the reference years industry was the greatest consumer of all types of energy. The share industry had in the total consumption of each energy source is shown in Table 2.

Table 2. Internal net final consumption of energy in Western Europe by type and by consuming sector in % of total consumption of each type for a number of reference years

Sector	Year	Solid fuels	Gaseous fuels	Liquid fuels	Electricity	Total
Industry	1950	42	67	29	63	43
	1955	41	74	32	63	44
	1960	43	77	34	62	46
	1965	44	72	36	57	44
	1970	54	64	33	55	44
of which iron-and-steel industry	1950	(12)	(51)	(4)	(10)	(14)
	1955	(12)	(51)	(5)	(10)	(15)
	1960	(15)	(48)	(5)	(10)	(16)
	1965	(18)	(38)	(5)	(10)	(14)
	1970	(27)	(29)	(4)	(10)	(13)
of which chemical industry	1950	n.a.	n.a.	n.a.	n.a.	n.a.
	1955	n.a.	n.a.	n.a.	n.a.	n.a.
	1961	(9)	(10)	(6)	(12)	(8)
	1965	(10)	(11)	(7)	(12)	(9)
	1970	(12)	(15)	(11)	(11)	(12)

For industry as a whole, the picture has been very stable. The share of the iron and steel industry in the consumption of coal shows a marked increase, due largely to the decreasing consumption by other sectors. The opposite is true of the iron and steel industry's consumption of gaseous fuels. The chemical sector increased its share for all sources considerably; in view of the total quantitative importance of oil, the rise, in oil consumption is particularly remarkable, but the consumption of gas by the chemical industry also increased quite sharply. In terms of total energy consumption, by 1970 the chemical sector had become as important as the iron and steel sector.

The distribution of the total energy consumption by each sector among energy sources is a better indicator of the relation between each industrial sector and each type of energy production. In Table 3 the shares in question are reproduced for all reference years. This table clearly shows that in 1970 solid

Table 3. Internal final consumption of energy in Western Europe by type and consuming sector, in % of total sectoral consumption

Sector	Year	Solid fuels	Gaseous fuels	Liquid fuels	Elec-ticity	Total
total industry	1950	69	13	10	8	100
	1955	58	17	15	10	100
	1960	44	20	24	12	100
	1965	30	19	38	13	100
	1970	22	22	42	14	100
iron-and-steel industry	1950	61	31	4	4	100
	1955	52	36	7	5	100
	1960	47	37	10	6	100
	1965	40	34	19	7	100
	1970	37	35	19	9	100
chemical industry	1950	n.a.	n.a.	n.a.	n.a.	100
	1955	n.a.	n.a.	n.a.	n.a.	100
	1961	50	14	23	13	100
	1965	36	14	36	14	100
	1970	17	20	53	10	100

fuel, though decreasing in importance, was still the major source of energy for the iron and steel industry, but that the relative share of oil had greatly increased. For the chemical industry the tendencies were much more pronounced: coal had, by 1970, lost its dominant place to oil. Grosso modo the conclusion from Table 3 must be that, while steel is still largely coal-oriented, chemicals have become predominantly oil-oriented, with gaseous fuels on the increase as well.

The iron-and-steel and chemical industries have been chosen as principal industrial consumption sectors, because for these categories statistics were readily available. That is not to say, however, that there might not be other important industrial users of energy. The OECD publications give no information on that score, but the EC has published statistical information in sectoral details covering a number of Western European countries. That information, somewhat rearranged, is reproduced in Table 4.

Table 4. Energy consumption by industry in 1965 in European Community (6 members) in million ton coal equivalent and in %

Industry	million ton coal equivalent	per cent
Iron and steel	76	29
Chemicals	61	24
Stone, glass, ceramics	35	14
Metal working	26	10
Food, beverages, tobacco	17	7
Textiles, wearing apparel, leather	13	5
Paper	13	5
Non-ferrous basic metals	11	4
Other	7	2
Total	259	100

Source: EC 1970.

The figures on the relative share of each sector in total consumption may have been influenced by the way the industries have been grouped into sectors, but the picture quite unequivocally indicates the iron-and-steel industry and the chemical industry as by far the largest energy consumers. That is evidence enough to consider the preliminary selection of the OECD statistics referred to above quite satisfactory.

A fairly recent development within the chemical industry is the increasing use of oil by the petrochemical industry for non-energy purposes. In 1950 and 1955 the production of oil products in petrochemical plants was still insignificant. In 1960, two per cent of total refinery production went into feedstocks, i.e. one third of the 6-percent share the chemical industry had in that year in total oil consumption; by 1970 that proportion had risen to 9 per cent, i.e. three quarters of the 12-percent share of the chemical industry in total oil production.

Within the iron-and-steel industry, blast furnaces take up the lion's share of energy, 66 per cent according to an EC study (EC 1966), while rolling mills account for 19 per cent, and steel mills for 12 per cent.

It appears, then, that the blast-furnace sector has a particularly close and important relation with the coal industry, and that the petrochemical industry has a similar relation with the oil industry. Both will be analysed further in the next section. It follows that within the energy sector itself, coke ovens and oil refineries would qualify for analysis, but coke ovens being much reduced in importance nowadays, the analysis will be confined to oil refineries along with petrochemical industries and iron-and-steel manufacture.

III OIL REFINING

The location of oil refining has always basically been determined by technological factors, although factors of a political nature have played a role as well.

Technologically, oil refining is a relatively simple, capital-intensive production process, by which one raw material, crude oil, is transformed, mainly by distillation, into a number of products. Most - in fact, until 1970 virtually all - sources of crude oil were located at some distance from the markets of their products in Western Europe. Before the Second World War refineries were located near the sources; in Western Europe hardly any refining was done. Since the war, however, oil refineries have sprung up all over the Western European market area. For one thing, the strategic importance of a continuous supply had come to be recognized; for another, balance-of-payment considerations played their part. The tendency towards market-orientation was then encouraged by the enormous increase in the demand for oil products; in fact, given the minimum scale a refinery requires from an economic and technical point of view, the present proliferation of refineries would not have been possible but for that increase in demand.

Still, all these factors together might not have contrived to bring about a major change, had they not been completed by another cause: the gradual extension of the uses to which the various refinery products could economically be put. For it was that development which in due course made it cheaper to transport in bulk one single commodity than to transport a large number of products; it was that tendency, indeed, which has led to complete market orientation after the war (Odell 1963, Frankel and Newton 1961, Butler 1963).

It is interesting to find out to what extent market orientation can also be observed on a smaller spatial scale within Western Europe. To that end some national figures on oil-product consumption and refinery capacity in the last two decades will be analysed. With the help of these figures, association indices have been calculated for the years 1950, 1955, 1960, 1965, and 1970. These indices measure the degree to which the distribution of refining capacity follows the distribution of the market for petroleum products. The mathematical expression of such an index is:

$$AI = 100 - 1/2 \sum_{j=1}^{n} \left\{ \frac{P_j \times 100}{\sum_{j=1}^{n} P_j} - \frac{M_j \times 100}{\sum_{j=1}^{n} M_j} \right\},$$

where AI is association index, P is production, M is market, and j is country. An association index expresses in one single figure to what degree the refining capacity of a certain area is adapted to the size of the areal markets for petroleum products. In the extreme case that in every country under observation the refining capacity is equal to the size of the market, the index is 100. In the other extreme, when the entire refinery industry is located in countries where there is no apparent consumption, and no refining capacity is located in any consuming country, the index will be 0.

Population and Gross National Product may serve as market indicators as well, as it is an observed fact that oil consumption is positively related to them. Association indices with respect to refinery capacity and these two market indicators have also been calculated. The results with reference to all three indicators are summarized in Table 5.

Table 5. Association indices of oil-refining capacity and three market indicators

Refining capacity	Market indicators	1950	1955	1960	1965	1970
Crude oil intake of refineries	Consumption of petroleum products	71	80	84	87	88
	Gross National Product	72	79	83	85	84
	Population	67	73	81	95	87

Sources: OECD - c, OECD - d, UN , Streicher, 1971.

All three indices show that the refinery industry tends to locate in the national market for its products. In recent years there have been indications, however, of the trend's losing impetus or even reverting.

Many authors have pointed out that with the growth of markets, refineries have become more market-oriented not only on a national, but also on a regional scale. That is difficult to verify, as regional-consumption figures are not available for all parts of the study area. Given the geographical structure of Western Europe, some indication of regional market location can be found in the shares of sea-coast and inland locations in refinery capacity. Table 6 visualizes the growing importance of inland locations.

Table 6. Additional refinery capacity in Western Europe by period and geographical area (in %)

Area	1950/1955	1955/1960	1960/1965	1965/1970
coastal	84	71	67	70
inland	16	29	33	30
total	100	100	100	100

Source: Streicher, 1971.

The figures relating to the period 1965/1970 seem to suggest a tendency to return to the sea coast. There are, indeed, several factors favouring such a shift. First, the refinery process is, as pointed out earlier, to a large extent technologically determined. It happens that the composition of regional demand does not match the range of products offered by the regional refinery. Seasonal variations and even completely random factors may widen the qualitative gap between supply and demand, making regional imports and exports necessary. Obviously, surplus production can be disposed of easier, at lower cost and to a wider range of destinations from port-based refineries than from land-locked ones, which puts the former at an advantage (Frankel and Newton 1968, Streicher 1971). Second, processing raw material at the break-of-bulk point necessitates only one handling, thus reducing costs. The construction of pipelines for transporting refinery products has reinforced the back-to-the-coast tendency, as has the development of energy-intensive industries and petro-chemical industries at sea-coast locations.

In order to explain the location mechanics of oil refineries, a simple model will be constructed, based on the transport cost and economies of scale, supposedly the most important factors influencing the location of the industry. There is a certain relation between the cost of transporting the products of a refinery and the economies of scale of the production process. Increasing the production capacity cuts down the cost of production, but causes distribution costs to rise, as products will have to be conveyed across a larger area. On certain assumptions about the spatial distribution of demand, an optimum refinery capacity can be worked out on the basis of the cost factors mentioned. In the classic example of a circular market area with uniform demand distribution, the problem to be solved can be stated as: for what level of capacity is the sum of production costs and costs of product transportation at a minimum, raw material being supplied at equal cost to all locations? In the Appendix it is described how that problem has been tackled by means of two functions, one to express the relation between cost of refining and refinery capacity, and a complementary one to express transportation costs in terms of refinery capacity (with the help of a variable representing the consumption of refinery products per areal unit). By minimizing the sum of the two functions the optimum refinery capacity can be found and studied further.

Two conclusions seem to suggest themselves from the results of the analysis:

1. The optimum capacity increases with increasing density of consumption.
2. The optimum capacity increases less fast than the density.

The latter conclusion implies that growing demand will be satisfied by additional production units. In reality, however, increased demand has been met in the last decade by extension of the capacity per unit rather than by their number. It must be assumed, therefore, that the reasoning developed above is too simple. Evidently, there are other factors at work, and among them the difference in the cost of crude oil should be mentioned first.

The commonest way to transport crude oil overland is by pipeline. Owing to the high initial investment required, transportation costs vary considerably for different sizes of refineries. It costs more to convey oil to many, relatively small and widely dispersed refineries than to only a few large units. When the costs of the transportation of crude oil are included in the analyses, a higher optimum capacity than the one resulting from the calculations presented above will be found. A modification of the second tentative conclusion imposes itself, therefore: the tendency to meet increasing demand by establishing new refinery units becomes less pronounced as soon as differences in input costs are taken into account.

More definite conclusions as to the effect of growing demand on the size and number of refineries could be arrived at if still other locational factors were included in the investigation. Within the scope of the present study that could not be done, however.

IV PETROCHEMICALS

The manufacture of chemicals from petroleum was pioneered in the United States during the inter-war years. At the time the organic chemical industry of Western Europe was still based on coal. Even after the Second World War, Western European organic chemical industry remained coal-oriented, using mainly coke-oven gases, coal tar and raw benzene, by-products of the coke production, as raw materials. Their output was, in fact, largely determined by the demand for the main product, coke, which came for the greater part from the iron-and-steel industry.

Two developments have greatly stimulated the growth of the petrochemical industry in Western Europe: (1) the relatively slow growth of the need for coke of the iron-and-steel industry, and (2) the extremely fast growth of the demand for organic chemicals (Guglielmo 1958, Wever 1974). When coke production no longer yielded sufficient by-products to satisfy the demand for chemicals, the gap was quickly filled by the petrochemical industry, which could use to an increasing degree the by-products (and the main products, for that matter) of the rapidly developing oil-refinery industry. Moreover, for some products petrochemical processes proved more efficient than carbochemical processes, so that the substitution of oil for coal gathered momentum during the 'fifties and 'sixties. Owing to the lack of comprehensive statistical material the development cannot be traced back to the 'fifties; some indications are available, however, such as the following statement in an OECD Chemical Industry report about 1955: "In general organic chemicals are manufactured in Europe by coal carbonisation, though the importance of petroleum as a basis for manufacture is rising." The rise was particularly rapid during the years 1955-1960, when the total production of petrochemicals rose from 0.4 to 1.7 million tons of carbon content. (OECD-e).

From publications issued by the OECD since 1962, figures can be borrowed relating to the total production of primary organic chemicals, by source of raw material, in the five most important producing countries (Western Germany, France, Italy, the Netherlands, and the United Kingdom, together accounting for about 90 per cent of Western European production), and for the years 1962, 1965 and 1970. These figures are given in Table 7. The figures in the table refer to primary organic chemicals.

Table 7. Raw materials of the basic organic chemicals industry in Western Europe

Year	Coal	Oil/Gas	Total
1962	38	62	100
1965	25	75	100
1970	7	93	100

Sources: OECD - e.

The figures clearly show how production has switched from coal to oil and natural gas as major raw material.

Since the present chapter is concerned with the petrochemical industry, it makes sense to delineate that industry within the chemical sector. this sector can be divided by type of product into the following groups of industries (OECD-e):

1. basic organic chemicals
2. basic inorganic chemicals
3. fertilizers
4. dyestuffs
5. plastics
6. paints and varnishes
7. soap and detergent
8. other.

The basic organic chemicals (group 1) are (EC 1970, Hahn 1970): synthetic gas for ammonia production, methanol, acetylene, olefins, and aromates. All these chemicals can be made from coal, oil, or natural gas; they serve as raw materials for the product groups 3 to 8. In this article, only plants that transform oil into "basic organic chemicals" will be referred to as "petrochemical industries". Oil refineries will not be included in the definition; basic organic chemicals not made from oil as well as the product groups 2 to 8 will be termed "other chemical industries". Ammonia, being made almost exclusively from natural gas or coal in

Western Europe, will not be considered any further, no more than the quantitatively unimportant products methanol and acetylene. That leaves for further analysis the group of olefins, consisting of ethylene, propylene, and C_4-olefins, and the group of aromates, made up mainly of benezene, toluene, and xylenes, which can all of them be made from a wide range of oil-refinery products. In practice, naphta is the most commonly used raw material; in 1965, approximately 70 per cent of all feedstocks for EC petrochemical industries was naphta. (EC 1970).

Naphta can be either cracked or transformed, to produce different ranges of chemical products. Table 8 indicates schematically what products in what

Table 8. Some technological aspects of the petrochemical industries

Process	Products	Yield %	Delivered to
Naphta cracking	Misc. gases & residues	35	Refinery
	Olefins		
	ethylene	25	Refinery and other chemical industries
	propylene	15	
	C_4-olefins	12	
	Aromates		
	benzene	5	Refinery (mainly for gasoline), and other chemical industries
	toluene	5	
	xylene	3	
	Total	100	
Naphta trans-form-ation	Aromates		
	benzene	8	Refinery (mainly for gasoline), and other chemical industries
	toluene	16	
	xylene	24	
	Miscellaneous residues	52	Refinery
	Total	100	

Source: EC 1970.

proportions (based on the amount of heavy naphta used) are yielded, and into what productions processes they are fed as inputs. Many end up as "plastics", but some of them are fed back to the refinery to be blended with other refinery products. Evidently, the petrochemical industry has a double link with the refineries, and location near them seems logical. Still, there may be other important location factors, and an analysis of the actual location pattern is in order, therefore. To our knowledge, the location of the petrochemical industry in Western Europe in after-year wars has not been comprehensively studied. An early work by Guglielmo (1958) does not refer specifically to the part of the industry established in Western Europe. Two geographers, however, have together covered a large part of the study field. Wever (1974) has analysed the 1970 situation for a number of petrochemical processes in the six countries of the old European Community; Chapman (1970, 1973) has studied the evolution of a number of petrochemical complexes (to the exclusion of untied industries) in the United Kingdom in the post-war period. For the year 1970 their combined analyses produce enough material for an analysis of the location of petrochemical industry in the seven countries of Western Europe that together account for over 90 per cent of the Western European production of basic organic chemicals.

In his analysis, Wever only takes the following two sequences into consideration: (1) naphta cracking → olefins → olefin products, and (2) naphta transformation → aromates → aromate products. The end-of-sequence industries, which are not sharply delineated, are also considered as petrochemical industries by Wever. Wever's theoretical considerations lead him to expect naphta-cracking installations in the proximity of refineries, and indeed, in 1970 all but four naphta crackers in the six EC countries were found to be linked to refineries by short pipelines. As far as naphta-transforming installations are concerned, Wever again concludes on theoretical grounds, mainly concerned with transportation costs and economies of scale, to a location near oil refineries. Indeed in 1970 almost all aromate-<u>producing</u> units were found at locations in the immediate vicinity of oil refineries,though aromate-<u>processing</u> industries were not.

Chapman's description of United Kingdom complexes do not contradict Wever's findings in the six EC countries. Both Wever's and Chapman's studies have also revealed, however, that the mere existence of a refinery somewhere is not enough to entice petrochemical works to lcoate there; in general, that will happen only when that refinery is located in the market for chemical products.

V IRON AND STEEL

Historically, the iron-and-steel industry used to locate in places where both iron ore and forests were found: iron ore as raw material, and forests to supply wood, to be used as an energy source in the form of charcoal. At that time transportation of either the ore or the wood over any distance was practically impossible. When about 1760 the technique of smelting with coke became generally known, the rapidly expanding iron industry shifted to areas where coal and iron ore were to be found (Roepke 1956, Reuss a.o. 1960, Pounds 1957, Steinberg 1967). Location continued to be determined by the cost of transporting energy and raw material. With improved sea transport and the introduction of railways came another shift in location: now iron- and-steel industries became freer in their place of settlement. Moreover, as Isard has pointed out (Isard and Capron 1949), advances in iron-and-steel technology had resulted in fuel economy, thus attenuating the pull of coal sites, and paving the way for market-oriented location. The increasing use of scrap, making it possible to produce steel with less iron ore and less energy, worked in the same direction, because by locating near the market, which is where most scrap is produced, the costs of transportation of scrap to the production site are minimized too (Isard and Cumberland 1950).

By the time the steel industry had grown relatively independent from its input orientation and free to become more market-oriented, the technology of steelmaking, in particular the wish to economize on fuel, had resulted in highly integrated iron and steel works with fairly large capacities. Isard and Kuenne (1953) conclude to the location of modern integrated steelworks with associated semi-finishing mills as close to the market as is permitted by site conditions, transportation facilities and socio-cultural factors. Given the geographical and political split-up of Western Europe,national states will often be the appropriate markets for the integrated, large-capacity steelworks of modern times. The orientation towards these national markets thus expected on economic and technological grounds, coincides with the tendency to locate steelworks in such a way as to further self-sufficiency of the states, for strategic reasons, or, with small countries, as an expression of new nationalism (Pounds 1959).

To what extent have combined strategic, technological and economic factors indeed led to market-oriented location of iron-and-steel works in Western Europe? To find that out, association indices have been calculated for the benchmark years 1950, 1955, 1960, 1965, and 1970, to represent the relation between the production of pig iron (from blast furnaces) and that of crude steel (from steel mills), and three variables that are commonly considered good indicators of the steel market, viz. metal-working industries, gross national product, and population. The results appear in Table 9; unfortunately figures on metal-working were available only for 1955, 1960, and 1965.

Table 9. Association indices: pig-iron production vs market indicators, and steel production vs market indicators, in Western Europe

Production of	Market Indicators	1950	1955	1960	1965	1970
Pig iron	metal working	n.a.	73	78	85	n.a.
	Gross Nat.Product	76	76	78	81	83
	population	64	66	68	73	75
Crude steel	metal working	n.a.	81	85	86	n.a.
	Gross Nat.Product	76	82	82	84	86
	population	66	70	72	76	80

Sources: OECD, Industrial Statistics 1900-1962, UN - , OECD - f, -d, OECD, The Engineering Industry 1962-1966, p. 26.

The consistent upward trend in the association indices points to an increasing orientation, for both kinds of iron-and-steel works distinguished here, towards national markets. It has already been pointed out that orientation towards regional markets would be difficult in the face of diseconomies of scale, which would rather favour plant specialization (Heal 1974). Still, market orientation could be encouraged in view of the increasing quality requirements of steel-consuming industries, which make frequent contacts necessary, while the availability of large quantities of scrap in a regional market, enabling small scrap-based steel industries to compete sucessfully with large integrated steelworks, may well reinforce again the tendency to locate in the regional market.

An analysis of the situation in Western Germany, a major steel-producing country with large private industries, may be elucidating here. A comparison made by Junius (1962) showed that North-Sea port locations had the advantage over inland locations in point of raw-material costs, but that the advantage shifted to Duisburg, Bremen and Dortmund as soon as the cost of transport to the market was taken into account. Locations farther inland were at a disadvantage in both cases. Evidently the Ruhr Area remains a good area for competitive steel making.

Association indices were once more put to play to detect any growing orientation towards a regional market in Western Germany. The indices were evolved from figures for 11 Bundesländer for the years 1950 and 1957 (the Saar not being included), and 1961, 1966, and 1970 (the Saar being included). Table 10 gives the

Table 10. Association indices of the iron-and-steel industry in Germany (FR) with some market indicators (11 Länder)

	Market indicators	1950	1957	1961	1966	1970
Iron-and-steel industry	metal-working industries	50	47	46	47	46
	Gross Domestic Product	50	50	48	48	48
	population	45	46	46	47	47

Sources: SBA - a, b, c, d.

indices; as could be expected, they are on a lower plane than those in Table 9, which referred to whole countries. They do show a stable market orientation throughout the twenty-year period observed. An analysis of the detailed tables of the individual Bundesländer confirms Junius's findings in that Nordrhein-Westfalen (the Ruhr Area) shows a constant, very high index (approximately 70); Rhein-Pfalz, an old inland raw-material-based location for the iron-and-steel industry, has lost ground to coastal Länder.

The conclusion must be that steel production is clearly market-oriented on a national level; location within the national market is determined by the access to regional markets and to raw materials. Thus, the location of steel industry remains a classical Weberian problem, and Van der Rijst (1969) undertook to explain the location of Western European steel industries with the help of a Weber-type calculation model. He concluded that much could be explained by the model, but that such factors as land, labour, water, and policy influences, not included in the Weber model, are becoming increasingly important. Environmental factors were not mentioned by Van der Rijst.

To recapitulate: technological changes have made the iron-and-steel industry shift, first from ore-and-forest areas to ore-and-coal areas; with thriftier techniques becoming available, the location became more independent from the source-of-energy stocks; better means of transportation made it possible to locate right away from the deposits of raw materials and sources of energy, in the market. These tendencies have been statistically confirmed. In the present situation, there are a number of other factors also to be taken into account.

VI CONCLUSION

In the foregoing sections it has been briefly analysed how the location of three energy-related industries developed in the period 1950-1970. The following conclusions can be drawn.

1. Refineries are market-oriented within the European space. On the country level they appear to have a certain preference for coastal locations. There seems no inclination to establish refineries near the place where the crude oil is produced.
2. The petrochemical industry is supply-oriented to the extent that, grosso modo, plants are located close to refineries, but not to the extent that there are petrochemical industries near every refinery; it is also oriented towards the market for its products, that is to say, towards other chemical industries.
3. The steel industry is market-oriented within the European space, but less so on the regional level, as could notably be established for the German situation.

There are no traces any more of orientation towards energy deposits.

One final remark suggests itself. Since 1970, two new factors that may affect the location of energy-related industries have entered upon the scene: the growing environmental concern, and the so-called energy-crisis. Further research will be necessary to find out their impact on the picture that has emerged in this paper.

REFERENCES

Butler, J.D., 1953, "The influence of economic factors on the location of oil refineries", Journal of Industrial Economics, Vol. I.

Chapman, K., 1970, "Oil-based industrial complexes in the U.K.", Tijdschrift voor Economische en Sociale Geografie, March-April.

Chapman, K., 1973, "Agglomeration and linkage in the U.K. petrochemical industry", Transactions of the Institute of British Geographers, Nov.

E.C. 1966, "De Economicsche invloed van de energieprijs", Series Economische en financiële zaken, Brussels.

E.G. 1970, "Etudes et enquêtes statistiques", nr. 4, Luxemburg.

Frankel, F.H. and Newton, W.L., 1961, "The location of refineries", Institute of Petroleum Review, Vol. 15, no. 175.

Frankel, F.H. and Newton, W.L., 1968, "Economies of Petroleum Refining, Present state and future prospects", Journal of the Institute of Petroleum, Vol. 54, no. 530.

Guglielmo, R., 1958, "La pétrochimie dans le monde", Paris.

Hahn, A. v., 1970, "The petrochemical industry, Market and Economics", New York.

Heal, D.W., 1974, "The steel industry in post war Britain", London.

Isard, W., 1948, "Some locational factors in the iron and steel industry since the early nineteenth century", Journal of Political Economy, Vol. 56, pp. 203-217.

Isard, W. and Capron, W.H., 1949, "The future locational pattern of the iron and steel production in the U.S.", Journal of Political Economy, Vol. 57, pp. 118-133.

Isard, W. and Cumberland, J.H., 1950, "New England as a possible location for an integrated iron and steel works", Economic Geography, Vol. 26, pp. 245-259.

Isard, W. and Kuenne, E., 1953, "The impact of steel upon the greater New York Philadelphia industrial region", Review of Economics and Statistics, Vol. 35, pp. 289-301.

Junius, H.P., 1962, "Zur Frage des Standortes neuzeitlicher Hüttenwerke in der BRD unter besonderer Berücksichtigung der Absatzorientierung", Aachen.

Odell, P.R., 1963, "An economic geography of oil", London.

OECD, 1964, "Industrial Statistics 1900-1962", Paris.

OECD, 1966, "Energy Policy", Paris.

OECD, 1968, "The Engineering Industry 1962-1966", Paris.

OECD, 1973, "Oil", Paris.

OECD - a, "Basic Statistics of Energy", Paris.

OECD - b, "Energy Statistics", Paris.

OECD - c, "Oil Statistics", Paris.

OECD - d, "National Account Statistics", Paris.

OECD - e, "The Chemical Industry", Paris.

OECD - f, "The Iron and Steel Industry", Paris.

Pounds, H.J.G., 1957, "Historical geography of the iron and steel industry in France", Annals of the Association of American Geographers, Vol. 47, no. 1.

Pounds, H.J.G., 1959, "The Geography of iron and steel", London.

Reuss, C., Koutny, E., and Tychon, L., 1960, "Le progrès économique en sidérurgie 1830-1955", Louvain.

Rijst, A. van der, 1969, "Beschouwingen over de vestigingsplaats van de Westeuropese Staalindustrie", IJmuiden.

Roepke, G., 1956, "Movements of the British iron and steel industry 1720-1950", Urbana.

SBA - a, "Statistische Jahrbücher", Wiesbaden.

SBA - b, "Das Bruttoinlandsprodukt der kreisfreien Städte", Wiesbaden.

SBA - c, "Die Nichtlandwirtschaftliche Arbeitsstättenzählungen", Wiesbaden.

Steinberg, A.G., 1967, "Die Entwicklung des Ruhrgebietes", Dusseldorf.

Streicher, H., 1961, "Mineralverarbeitung in West Europa", Hamburg UN, "Demographic yearbooks", New York.

Wever, E., 1974, "Raffinaderij en petrochemische industrie, ontstaan, samenstalling en voorkomen van petrochemische complexon", Nijmegen.

APPENDIX

ECONOMIES OF SCALE AND TRANSPORTATION COSTS

It is assumed that three factors determine the optimum capacity of refineries: economies of scale in the production process, costs of transporting the products, and distribution of demand.

Regarding the costs of production it is assumed that the following relation holds:

$$k = \alpha(c - \beta)^{-1} + \qquad (1)$$

k: costs in guilders per ton

c: capacity in mln tons per year

Regarding the costs of transporting the products the following relation is assumed:

$$t = \delta d + \Psi \qquad (2)$$

t: transportation costs in guilders per ton

d: distance in km.

When we consider the hypothetical case of a circular market with radius r and uniform distribution of demand per areal unit, the average distance the products have to travel from the centre is 2/3 r. For this case the average transportation costs per ton of product are:

$$t = 2/3\ \delta r + \Psi \qquad (3)$$

Now we want to express t as a function of the capacity of a refinery of which the products are distributed in a uniform way over a circular area. The capacity of such a refinery is:

$$c = a\pi r^2$$

c: capacity in mln tons per year

a: demand in mln tons per km^2

or

$$r = (a\pi)^{-1/2} c^{1/2} \qquad (4)$$

and

$$t = 2/3\ \delta (a\pi)^{-1/2} c^{1/2} + \Psi \qquad (5)$$

Costs of transportation (t) and costs of production are now expressed in the same variable: refinery capacity (c).

Total costs are now:

$$K = \alpha(c-\beta)^{-1} + \gamma + \rho c^{1/2} + \Psi$$

and these are at a minimum when

$$\frac{dk}{dc} = 0$$

or

$$-\alpha(c-\rho)^{-2} + 1/2\ \rho c^{-1/2} = 0$$

or

$$\alpha(c-\rho)^{-2} = 1/2\rho c^{-1/2}$$

The value of c for which this equality holds has graphically been determined for some values of the parameters. The values are shown in figure 1, in which the intersections are optimum capacities, for which the correspondence with density of demand is:

C opt. mln tons/yr	d tons/km^2
9.9	150
12.5	440
13.5	600
16.8	1400
20.2	3180

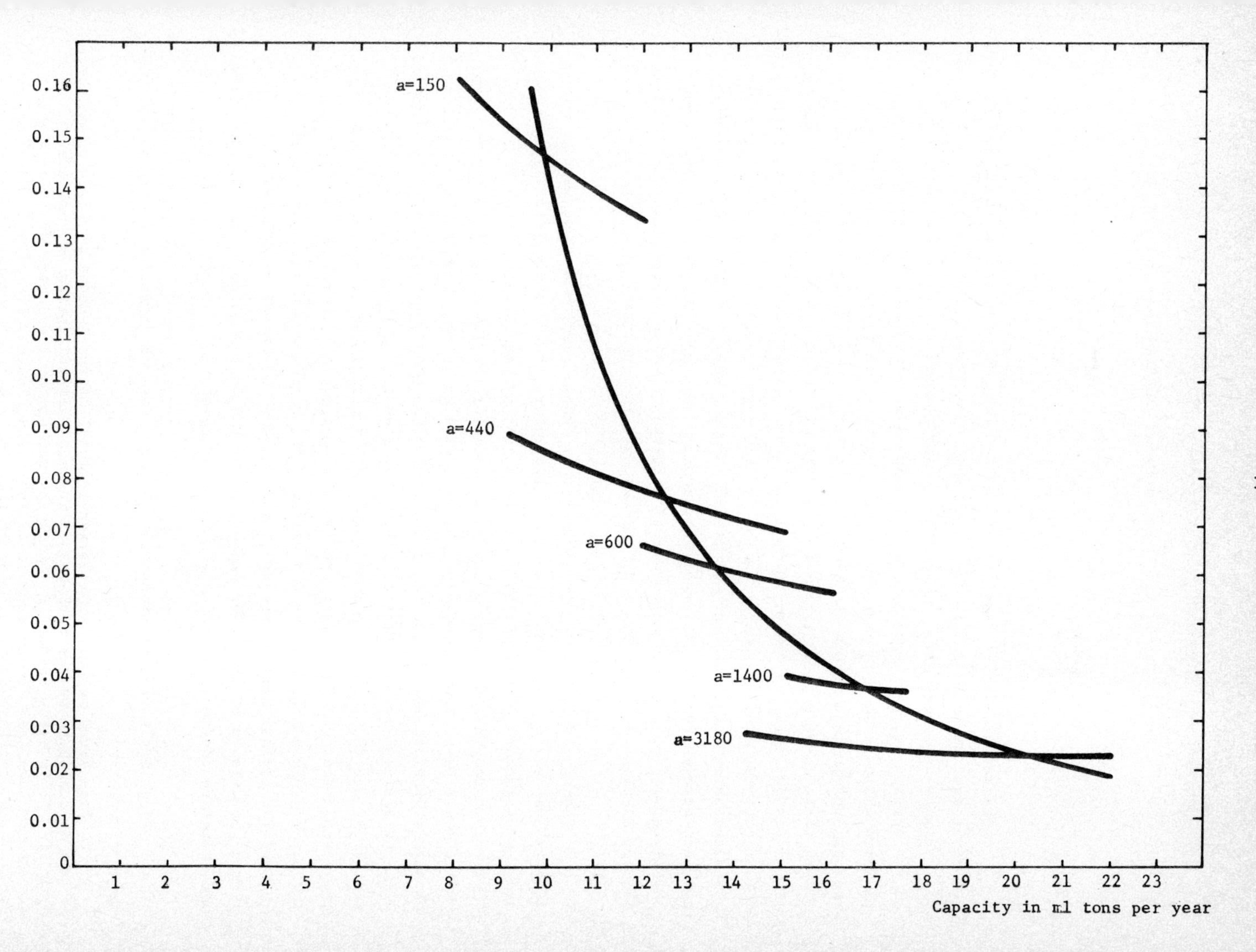

Molle and Breimer, figure 1.

REGIONAL BALANCED GROWTH IN ITALY AND THE INCREASE IN OIL PRICES[1]

Murray Brown, Maurizio Di Palma and Umberto Triulzi

I INTRODUCTION

The recent rise in fuel prices has had many diverse consequences both for the world economy and for domestic economies. Notwithstanding the fact that aggregate demands have been affected and that major attention is currently being focused on the inflationary and employment difficulties experienced by many economies, it is worth while to enquire into the effect that the oil price rise may have on regional development within an economy. The purpose of the present paper is precisely that, namely, to determine whether the oil price rise, itself, is a balancing or disequilibrating factor in regional development. Specifically, we put this question to the Italian economy in the reference period, 1951 to 1974. Certainly, the accentuated development of the Italian economy over the last twenty years has not been accompanied by a balance in economic differences within the regions and the sectors. The regional and sectoral differences which characterised the Italian economy during the 1950's are still present and in some cases even more pronounced.

Planning policy since the 1950's (the Vanoni plan) and subsequently more detailed reports of the 1960's (the La Malfa report, the Giolitti plan 1964-1969, and the Pieraccini plan 1965-1970, the 80's project, the Giolitti plan 1970/75) have always considered as among the priorities for the development of the country the elimination of this imbalance. It was believed in fact that the accumulation process of the productive system of the South (the poor regions) should be financed with the savings from the central North (the rich regions). During the planning period, it was thought, while awaiting this industrial take-off, the Southern regions would continue to satisfy the labor demand of the Central North where the labor force was scarce and the Central Northern region would send capital to the South. The redistribution process of the primary factors of production would permit the balance of the labor budget and of the capital account in the two areas, and therefore the realization of a more equal development process in the country.

Thus in the various productive sectors, the modernisation processes and the

---

[1] This study is part of a larger regional-national model of Italy which has as its central purpose the analysis of regional disequilibrium and policies to reduce it. It is being conducted jointly by members of the Centro di studi e piani economici, Rome, and the Department of Economics, State University of New York at Buffalo. A preliminary description of the complete model appears in Murray Brown, Maurizio Di Palma and Bruno Ferrar, (1972), pp. 25-44.

enlargement of the already existing structures, as well as new initiatives, were encouraged to reduce the differences in the efficiency levels within the various sectors by way of a planned policy of sectoral investments in the agricultural sector. This would lead to a large exodus of the labor force from this sector towards the industrial and tertiary sectors and also increase the efficiency of the agricultural sector, eliminating the existing "dualism".

How successful was this policy? To answer this question we can look at the behaviour of an index of regional development namely regional income, income per worker, which we shall discuss in more detail below. In Table 1, regional sectoral productivity for 1951, 1959 and 1968 and the weighted average productivities for the 19 regions are shown.

It is seen immediately that there was a continuous reduction from 1951 to 1968 (more accelerated in the Northern than in the Southern region), of the weight of the agriculture sector in the formation of regional gross product per worker. In fact, while for the Northern regions (Piemonte, Lombardia and Liguria) the weight of agriculture in 1968 is not more than 6 - 7% and for the Central North a little more than 10% (except Marche and Emilia-Romagna), for the Southern regions the contribution of agriculture to value-added per worker is generally over 20%. The notable reduction of the weight of this primary sector is counterbalanced by the very high increase in all regions of the tertiary sectors, i.e. services and transport, in terms of employment as well as productivity.

In 1951 the percentages of these sectors in total regional value added for the North and North-East regions were less than 20%, while in 1968 these values are close to 30%; in the central and southern regions the tertiary sector shows a higher increase. For most of these regions in fact it is possible to say that the weight of services comes out to be higher than that of the industry, with extreme values in Lazio (42,6%), Marche (39,1%), Toscana (37,8%) and Campania (34,2%).

With respect to industry it has declined within most regions in favour of other sectors, namely the construction sector and services in general. The only regions that shows an increase are Liguria (from 34,1% in 1951 to 35,2% in 1968), Emilia-Romagna (from 29,4% to 32,7%), Campania (from 24,5% to 26,8%) and Puglia (from 21,3% to 23,0%).

The construction sector shows very high rates of increase in regions like Liguria, Trentino-Alto Adige, Basilicata and Calabria, which did not receive significant migration flows or did not have high rates of industrialization, while in some others like Piemonte, Lombardia and Lazio, the productivity of the sector has not increased markedly, although there has been considerable industrialization and high rates of urbanization.

Let us discuss the major trend of behaviour of regional productivities from 1951 to 1968 as seen from Table 1. It turns out that the largest increases have been realized by the poorest regions of the South. Encouraging as that is, nevertheless such increases have not been sufficient to reduce significantly the economic differences existing among these regions. The average productivity for the richest regions of the North in 1968 is about 3 millions (of Italian lire) per worker, while that of the southern regions (except Sicilia and Sardegna) is only 1,6 million per worker. For the central regions there has been a consistent reduction of the "gap" with the North and in fact the levels of productivity, for the last year, are very near to those of the North. It is in this context of regional growth we wish to evaluate the effect of the oil price rise.

II INCOME PER WORKER BY REGION

We are concerned with the impact of the recent increase in fuel and energy prices on regional growth. Before specifying the question more precisely, we must indicate how we propose to measure regional growth. Notwithstanding the fact that there are several indices of regional growth, each of which has advantages and disadvantages (Kraft et.al. (1971)), we use an index of regional income (or value-added) originating with respect to the number of employees in the region. We denote the income per worker in region j, sector i, as $\lambda_{ij} = v_{ij}/l_{ij}$, where v_{ij} is value-added at current factor cost and l_{ij} is the number of workers employed in the ith sector and the jth region. Presumably, it would be preferable to measure regional growth in terms of income per capita; we shall see why this is not feasible in the present study.

Each region's value-added is produced by seven sectors (agriculture, industry, construction, commerce, transportation, housing and public administration). Since public administration is exogenous and energy requirement in housing is negligible, these sectors are omitted from our analysis. Adding the weighted value-added quantitites per worker over the sectors within a region, we have

$$\lambda_{\cdot j} = \sum_{i=1}^{5} x_{ij} \lambda_{ij} \qquad (1)$$

where $x_{ij} = l_{ij}/\sum_i l_{ij}$, i.e. the proportion of sector i's employment in total regional employment. Now, it is clear why the measure of regional growth is not taken in terms of value-added per capita, for data related to population dependent on different sectors are not available nor are they conceptually reasonalbe quantities. In any event, λ_{ij} and $\lambda_{\cdot j}$ are the numbers reported in Table 1.

III FUEL, CAPITAL AND LABOR IN THE PRODUCTION PROCESS

It is customarily thought that fuel or energy is combined in fixed proportions with capital in the production process. So in the short run and perhaps for a longer period - a rise in the price of fuel increases the cost of capital thus reducing the use of both capital and fuel relative to labor. To see this, let a typical neoclassical production fuction be

$$v = f\left[1, \min_{k,m}\left(\frac{k}{a}, \frac{m}{b}\right)\right] \tag{2}$$

where f is a constant returns to scale function, v is output, l is labor, k is capital, m is fuel and a and b are constants (the result does not change if we relax the constant returns assumption). Then, at the point where k/a = m/b, the marginal product of fuel is

$$f_m = \frac{\delta f}{\delta m} = \frac{\delta f}{\delta k}\,\frac{a}{b} = p_m \tag{3}$$

where p_m is the fuel price. Thus, a rise in the price of fuel reduces the capital-output ratio since

$$f_m = f_m\left(\frac{v}{k}\right)$$

and

$$f_m' > 0$$

Now, suppose that within a region, two connected sectors (i.e., one sector uses the output of the other, say, capital goods, as an input), operate with such a production function. Then, according to the Rybczynski theorem (Jones (1965)) the reduction in capital, holding all other parameters constant will decrease the output of the good using more fuel and increase the output of the good with less fuel. If the output of the first good is weighted more heavily in the region's total output and if prices do not change in proportion, then the value of output of the first region rises while that of the second falls. Thus, knowing the weights each output has in total regional value-added and the intensities in the use of capital in production in the various sectors of two regions, one can determine how the total regional value-added behaves as a result of the rise in the price of fuel. Comparisons of the two region's value-added movements are thus possible.

The problems inherent in this approach are that the international trade theorems to which we have just referred do not generalize easily to more than two sectors. One has to have considerably more knowledge of the production process than the simple factor intensities in order to draw the Samuelson-Stolper and Rybczynski implications. Moreover, the theory is presented in the context of a constant returns to scale, competitive economy, an unacceptable assumption in the case of the Italian economy (and perhaps others as well). An alternative way to treat the fuel price rise is adopted in the present paper. It has its own difficulties, but it can be implemented.

IV A METHOD FOR EVALUATING THE EFFECT OF FACTOR PRICE INCREASES ON REGIONAL INCOME PER WORKER

We continue to assume that fuel enters the production process in all sectors in roughly fixed proportions with capital. Moreover, we use the estimates of the regional value-added distribution equations for Italy derived in Brown (1968). The value-added in region j for i^{th} sector, i.e. v_{ij} can be shown to depend upon total regional value-added, $v_{.j}$, the sector's own price, p_{ij}, and the overall regional price, $p_{.j}$:

$$v_{ij} = B_{ijo} \, v_{.j}^{B_{ij1}} \, p_{ij}^{B_{ij2}} \, p_{.j}^{B_{ij3}} \tag{4}$$

In this form, the B_{ijh} (h = 1,2,3) are constant elasticities with respect to the relevant variables and they have the following sign properties:

$$B_{ij1}, B_{ij2} > 0; \quad B_{ij3} < 0.$$

The function is derived and its properties are discussed in the abovementioned paper so that we can proceed directly to indicate how the estimates of the function can be used to determine whether the rise in fuel prices have benefited or impeded regional growth.

Dividing (4) by l_{ij} and multiplying by x_{ij}, we have

$$x_{ij} \, \lambda_{ij} = c_{ij} \, p_{ij}^{B_{ij2}} \, p_{.j}^{B_{ij3}} \tag{5}$$

where

$$c_{ij} = B_{ijo} \, v_{.j}^{B_{ij1}} \, l_{ij}^{-1} \, x_{ij} \tag{6}$$

We shall assume that c_{ij}'s are constants. This mean that the estimates are applicable only to the short run since the effect of the rise of fuel prices is held constant on total regional output, the share of the sector's employment in total regional employment, and on the sector's own employment. These are serious limitations but until the effects on demand can be evaluated, we would follow this course.

Suppose now that fuel prices rise and that we can evaluate the effect on the own price and on regional price. In a study using an input-output approach, these effects have been evaluated for Italy.[2]

---

[2] Centro di studi e piani economici, (1974). The aim of this study was to evaluate the effects of oil price increases on the prices of (continued, next page)

Because an input-output matrix with fixed coefficients has been used, this is an additional reason for considering the present study to be a short run analysis. We denote the own price and regional price after the fuel price rise by barred variables, i.e., $\bar{p}_{ij}$ and $\bar{p}_{.j}$, respectively.

The difference in the weighted income per worker between the period after the fuel price rise and the base period is $x_{ij}\ (\bar{\lambda}_{ij} - \lambda_{ij})$, i.e.

$$x_{ij}\ (\bar{\lambda}_{ij} - \lambda_{ij}) = x_{ij}\ \lambda_{ij} \left[\left(\frac{\bar{p}_{ij}}{p_{ij}}\right)^{B_{ij2}} \cdot \left(\frac{\bar{p}_{.j}}{p_{.j}}\right)^{B_{ij2}} -1 \right] \tag{7}$$

Summing over the five sectors gives the change in total regional income per worker as a result of the fuel price rise:

$$\bar{\lambda}_{.j} - \lambda_{.j} = \sum_{i=1}^{5} x_{ij}\ \lambda_{ij} \left[\left(\frac{\bar{p}_{ij}}{p_{ij}}\right)^{B_{ij2}} \cdot \left(\frac{\bar{p}_{.j}}{p_{.j}}\right)^{B_{ij3}} - 1 \right] \tag{8}$$

This is the relationship we have evaluated for the effect of the change in fuel prices on regional income per worker.

Note that we have taken the summation of (8) over five sectors, omitting housing and public administration. This constitutes a further reason for the study being termed a short run analysis. We have also held constant the elasticities. Clearly, in order for the analysis to be extended to the long run, the restrictions responsible for the short run have to be relaxed.

V RESULTS OF THE SIMULATION

Table 2 gives the values for $(\bar{\lambda}_{.j} - \lambda_{.j})$, the change in total regional income per worker as a result of the fuel price rise for nineteen regions in Italy which is obtained from evaluating equation (8).

It is immediately clear that all regions are adversely affected by the oil price rise but the impact is by no means uniform over the regions. Toscana, Liguria and Piemonte are affected most while Lazio, Calabria, Basilicata, Lombardia and Veneto are least affected.

The principal question is: were poor regions less adversely affected by the oil price rise than the regions with high income per worker. As noted above,

---

2 (continued)
the various productive sectors of the Italian economy. In order to analyse these effects an input-output table, that of the Italian economy for 1969, has been used in which all interrelations between oil sectors and other sectors of the national economic system were considered.

we measure regional "well-being" by income per worker, $\lambda_{.j}$, which is given in Table 1. A ranking of the two appears in Table 3. The Spearman rank correlation coefficient between $\lambda_{.j}$ and the absolute value of $(\bar{\lambda}_{.j} - \lambda_{.j})$ for all nineteen regions is 0.60. Being positive and reasonably large, it appears that the oil price rise tended to affect poorer regions less than the richer regions.

An interesting result can be obtained by comparing income per worker $\lambda_{.j}$ and the absolute value of the impact of the oil price rise on income per worker, i.e. $\bar{\lambda}_{.j} - \lambda_{.j}$, for those regions in the first area, the northwest, and those the fourth area, the south plus the islands. The Spearman rank correlation is 0.32. This means that the oil price rise has not aided the south in catching up to the rich northwest as much as it aided all of the regions in Italy. In other words the oil price rise seems to have helped the north-east and central regions more than it has aided the south. It appears that, at least in the short run, the south has done poorly relative to the northeast and center, though it has benefited from the overall equilibrating effect of the oil price rise. Thus, the pattern we observed in our discussion of Table 1 has been preserved even with respect to the short-run effect of the oil price rise.

With respect to the impact of the oil price rise on the various sectors, we note the following results. Agriculture and construction appear to have been most adversely affected, the impact on industry is less, and on transportation and services it is the least. In the rich northwest, industry is affected more adversely than in the rest of the economy, as one would expect. Again, we note that Lombardia is an exception to this pattern. In the central part of Italy, construction is uniformly hit badly while in the south it is hardly affected. Except for a few regions, the oil price rise seems to have affected services and transportation hardly at all.

It is questionable whether these patterns will continue to persist in the long run. If they do, then our analysis will have been proved to be surprisingly robust. In any event, long and short-run policies to energy (*inter alia*) will have to take into account these results, especially if balanced regional growth retains its place as one of the principal social concerns.

REFERENCES

Brown, M., 1968, "The distribution of Sectoral Value-Added by Region in Italy, 1961-1968", Discussion Paper 318, Dep. of Economics, State University of New York at Buffalo.

Brown, M., Di Palma, M., Ferrara, B., 1972, "A regional-National Econometric Model of Italy", Papers of the Regional Science Association, Vol. 29.

Centro di studi e piani economici, 1974, "The effect on national economic structure determined by different variations in oil prices", Ente Nazionale Idrocarburi (ENI), Roma.

Jones, R.W., 1965, "The Structure of Simple General Equilibrium Models" in The Journal of Political Economy, LXXIII.

Kraft, G., Willens, A.R., Kalev, J.B., Meyer, J.R., 1971, "On the definition of a depressed area" in Kain, J.F., Meyer, J.R., (eds.), Essays in Regional Economics, Harvard University Press.

TABLE 1: Regional and Sectoral Income per Worker in Italy, 1951, 1959, 1968

	1951							1959						
Regioni	A%	I%	C%	S%	T%	Tot(000) di lire	%(2)	A%	I%	C%	S%	T%	Tot(000) di lire	%(2)
Piemonte	18.3	57.4	2.4	16.6	5.3	507	85.6	12.9	50.1	5.0	23.8	7.6	856	94.6
Valle d'Aosta	17.4	58.6	2.8	14.2	7.0	471	79.4	11.5	52.3	7.1	21.5	7.6	873	96.5
Liguria	19.3	34.1	6.7	27.3	12.6	388	65.4	8.8	42.7	9.9	27.6	11.0	889	98.2
Lombardia	13.2	62.8	2.4	16.9	4.7	593	100.0	10.2	50.5	6.6	27.2	5.5	905	100.0
Trentino-Alto Adige	29.4	46.5	4.3	13.7	6.1	439	74.0	22.5	41.5	7.5	21.0	7.5	691	76.4
Veneto	33.7	33.1	4.6	19.8	8.8	329	55.5	27.4	27.2	7.9	25.7	11.8	604	66.7
Friuli Venezia Giulia	23.2	41.0	4.3	20.4	11.1	324	54.6	13.8	40.6	8.1	25.7	11.8	712	78.6
Emilia-Romagna	39.5	29.4	5.5	19.0	6.6	347	58.5	30.7	27.5	9.0	26.3	6.5	691	76.4
Marche	40.0	29.0	5.3	19.6	6.1	245	41.3	30.3	27.2	6.3	29.2	7.0	442	48.8
Toscana	20.2	48.3	4.7	21.3	5.5	381	64.2	14.4	40.7	6.9	30.2	7.8	706	78.8
Umbria	43.2	24.4	4.7	21.9	6.0	234	39.5	25.9	34.6	5.6	26.5	7.4	514	56.8
Lazio	18.2	33.6	5.6	29.6	13.0	324	54.6	14.0	21.6	11.2	37.2	16.0	663	73.3
Campania	34.3	24.5	4.9	24.5	11.8	204	34.4	22.1	28.3	7.5	28.6	13.5	480	53.0
Abbruzzi) Molise)	48.6	20.3	6.8	17.5	6.8	177	29.8	35.0	25.6	9.4	24.3	5.1	403	44.5
Fuglia	45.9	21.3	5.7	20.9	6.2	211	35.6	37.7	17.6	9.2	27.5	8.0	448	49.5
Basilicata	51.1	18.7	9.3	16.5	4.4	182	30.7	43.2	12.7	8.6	26.6	8.9	315	34.8
Calabria	40.0	20.9	6.1	18.4	5.6	196	33.1	33.2	19.7	8.6	30.7	7.8	361	39.9
Sicilia	40.9	25.3	7.0	20.9	5.9	273	46.0	29.8	21.2	9.6	27.9	11.5	520	57.5
Sardegna	25.2	50.8	4.7	12.7	5.6	408	68.8	30.8	20.2	11.4	27.6	10.0	519	57.3

Source data: An econometric model for the national-regional development of Italy, Vol. II, It. Centro di Studi e Piani Economici, Roma, 1971. (1) The symbols A,I,C,S,T, are for Agriculture, Industry, Construction, Services, Transport. (2) Regional income per worker in Lombardia equals 100.

TABLE 1 (continued)

Regioni	1968							1968/1951
	A%	I%	C%	S%	T%	Tot(000) di lire	%(2)	Reg. income per w '68 / Reg. income per w '51
Piemonte	6.9	49.0	8.7	28.3	7.1	2.746	99.3	441.6
Valle d'Aosta	4.6	31.1	37.2	21.5	5.6	3.660	133.4	577.0
Liguria	5.2	35.2	17.0	30.8	11.7	3.490	126.3	800.0
Lombardia	6.2	46.1	7.7	34.0	6.0	2.764	100.0	366.1
Trentino-Alto Adige	10.6	31.8	21.5	28.4	7.7	2.320	83.9	428.5
Veneto	13.5	31.7	14.8	31.1	8.9	2.192	79.3	556.3
Friuli Venezia Giulia	7.5	34.0	14.6	32.3	11.6	2.485	89.9	567.0
Emilia-Romagne	14.5	32.4	13.0	32.8	7.0	2.264	81.9	552.4
Marche	17.0	22.4	13.5	39.1	8.0	1.655	59.9	575.5
Toscana	9.2	35.7	9.0	37.8	8.3	2.365	85.6	520.7
Umbria	13.6	34.2	9.4	35.7	7.1	2.014	72.9	760.7
Lazio	8.5	23.6	10.7	42.6	14.6	2.679	96.9	726.8
Campania	19.0	26.8	10.0	34.2	10.0	1.677	60.7	722.0
Abbruzzi) Molise)	25.7	17.4	16.3	31.7	8.9	1.547	56.0	774.0
Fuglia	25.9	23.0	10.1	31.8	9.2	1.564	56.6	641.2
Basilicata	20.4	17.2	30.5	26.3	5.6	1.591	57.6	774.2
Calabria	20.9	10.6	28.6	33.2	6.7	1.635	59.2	734.2
Sicilia	25.9	20.4	11.3	31.9	10.5	1.862	67.4	582.0
Sardegna	21.3	23.4	12.3	32.9	10.1	2.225	80.5	445.3

TABLE 2

Effect of Oil Price Rise on Income per Worker by Region (in Lire)[1]

REGIONS	
Piemonte	- 1.526.920
Valle d'Aosta	- 1.071.592
Liguria	- 1.681.283
Lombardia	- 149.780
Trentino-Alto Adige	- 824.753
Veneto	- 139.543
Friuli-Venezia Giulia	- 1.441.497
Emilia-Romagna	- 316.008
Marche	- 365.919
Toscana	- 2.116.947
Umbria	- 1.001.094
Lazio	- 301.487
Campania	- 368.396
Abruzzi e Molise	- 690.450
Puglia	- 362.107
Basilicata	- 158.897
Calabria	- 186.587
Sicilia	- 484.364
Sardegna	- 1.081.410

[1] Obtained by evaluating equation (3) as explained in the text.

TABLE 3

Ranking of Income per Worker and the Effect of the Oil Price Rise on Income per Worker by Region

INCOME PER WORKER	ABSOLUTE VALUE OF EFFECT OF OIL PRICE RISE
Valle d'Aosta	Toscana
Liguria	Liguria
Lombardia	Piemonte
Piemonte	Friuli-Venezia Giulia
Lazio	Sardegna
Friuli-Venezia Giulia	Valle d'Aosta
Toscana	Umbria
Trentino-Alto Adige	Trentino-Alto Adige
Emilia Romagna	Abruzzi e Molise
Sardegna	Sicilia
Veneto	Campania
Umbria	Marche
Sicilia	Puglia
Campania	Emilia-Romagna
Marche	Lazio
Calabria	Calabria
Basilicata	Basilicata
Puglia	Lombardia
Abruzzi e Molise	Veneto

A DYNAMIC REGIONAL ANALYSIS OF FACTORS AFFECTING THE ELECTRICAL ENERGY SECTOR IN THE U.S.

Rajendra K. Pachauri

I INTRODUCTION

The electrical energy sector in the U.S. economy is organized on the basis of a number of specific geographic regions, each served by an individual electric utility franchised and legally bound to supply electric power within the area. A study of the aggregate growth of the electrical energy sector within the nation must, therefore, necessarily be based on specific factors influencing each region as a unit of the whole. Most econometric models in this field have concentrated on representing the national aggregate only, and have generally not included the various specific feedback effects characterizing the growth of the electrical energy sector on a dynamic basis.

This paper discusses some aspects of a model constructed by the author representing a specific region in southeast U.S.A.,[1] dealing with the growth of electrical energy in the area. The essential basis for any such model is the formulation and estimation of relationships representing the demand for electrical energy. An electric utility determines its investment, construction and capital-raising plans on the basis of demand forecasts in the future. Such demand forecasts till quite recently were obtained with reasonable accuracy purely by extrapolating past data. The suitability of such "naive" techniques inhibited the development of more refined and robust forecasting methodologies. Further, errors in forecasting were not costly to a utility, since any gaps in available capacity and actual demand were made up by the quick installation of gas-fired internal combustion turbine generators or even fossil-fuel fired steam generating plants. This has changed with the shortage and price increases of hydrocarbon fuels, and with the development of nuclear power generation. Nuclear power can provide electricity at least costs. The demand for electricity which had been growing steadily at high rates in the past few years on account of steady growth in per capital income and a steady decline in real price of electricity in the U.S., has slowed down in recent years on account of changes in trends in these very factors. Data from the Edison Electric Institute's Year Book for 1973 enables us to calculate the ratio of electricity prices to the Consumer Price Index for each recent annual period. These ratios for the years 1965-73 were respectively 1.0487, 1.0195, 1.00, 0.9683, 0.9132,

---

[1]The region described covers parts of North and South Carolina and is precisely the region served by Carolina Power and Light Company.

0.9332, 0.9489, 0.9384. Price increases in 1974 were completely beyond the range suggested by these figures, and the outlook for the future is even bleaker.

One reason why the price elasticity of electrical energy demand has been underrated in academic and business cirlces is the presence of long lags in quantity response to price changes. Steady trends in electricity prices have obscured the visibility of responses in quantity demanded and a general awarenes of this price effect. Halvorsen (1972) estimated the long run elasticity of demand for the U.S. using a variety of lag structures and found little difference in the values of elasticity obtained from each model, indicating the high degree of autocorrelation in pre-1971 price and quantity data. He (1973c) found the long run elasticities of demand to be close to unity for the residential and commercial sectors, and between -1.752 and -1.530 for the industrial sector.

The supply of electricity is dependent on the capacity available with the utility concerned. Decisions on increasing the capacity of a system are based on future forecasts of (1) energy demand in kilowatt-hours, (2) the extent of the load in kilowatts and related variations and peaks in demand, (3) the availability of generating plant at any given time and allowances for maintenance schedules and outages. The utility planner will attempt to provide that portion of the required capacity by nuclear generation which represents a stable load, and the peaking or varying component of capacity by provision of other means which have lower capital costs. The capital cost of nuclear plants is currently estimated to be around $500.00/Kw and that of internal combusion turbine plants around $100.00/Kw; marginal costs of operation of these are, however, quite the reverse of capital costs. In a capital-intensive industry such as that supplying electric power, errors in capital planning can either leave the industry with deficiencies in capacity on the one hand or with excessively high costs of supplying power on the other. A proper forecast of the time path of electricity demand in the future is therefore essential to the financial and operational success of a supplier.

II STRUCTURE OF THE DYNAMIC MODEL

The demand side of the dynamic model described in this paper is based on microeconomic factors affecting the quantity of electricity demanded by each individual customer and macroeconomic-demographic factors determining the number of customers within the region. Total demand is shared between three separate sectors - residential, industrial and commercial. The residential sector includes three distinct rate groups: the first consists of customers with all-electric homes, the second of those with approved electric water heaters only and the third group of all other residential customers. The industrial sector in the model was broken down into seven major industry groups on the basis of the major industrial activities within the region modelled. These groups were formulated in respect of food and kindred products, textile mill products, apparel and other textile products, furniture and lumber, chemicals and machinery respectively and the balance of all other industries constituting the seventh group. The commercial sector typically in the U.S. consists of a large variety of rate groups. In our regional model we have aggregated these rate groups to form five major groups, which represent, as far as possible, homogeneous sets of activities. These are in the categories of business, schools and educational institutions, churches, commercial housing and housing construction respectively.

The purpose of this model being mainly to simulate the electrical energy sector over the long run future of the region, all those factors were considered in our specification that seemed subject to distinct change in the future. Within the electrical energy sector itself a considerable change in the mix of customers is likely to take place. Hence, it was essential to investigate the determinants of the level of customers within each customer group at any time. These included both national macroeconomic variables as well as factors specific to the economy of the region. The description below deals with these factors as well as microeconomic variables affecting each customer unit in the three major sectors.

III RESIDENTIAL SECTOR

The total number of households within a region determines the number of residential customers of electricity at a given time. Hence, a prediction of total population in the region requires estimation of birth rates, death rates and the rate of net migration. The total number of households is then determined by the product of average size of household multiplied by the total population existing. The proportion of customers distributed between the three residential rate groups would be determined by the average energy consumed within each and a time trend representing all other variables which cause changes in life style over time, and bring about an increase in electricity used in the home. The rationale for this assumption lies in the declining block structure of electric rates which leads to a reduction in price per Kwh with increases in quantity consumed. Hence, given a certain rate structure, the average Kwh per customer has an inverse relationship with average price. If the average Kwh for a rate group represents a function of customer trends and preferences, any change in this variable would, through an inverse price effect, act in attracting or repelling customers from other rate groups. The quantity demanded by each individual customer was specified as being dependent on quantity demanded in the previous period, the prices of electricity and natural gas, per capita income and a time trend representing the introduction of new appliances. Figure 1 shows diagrammatically the factors affecting quantity variables in the residential sector.

IV INDUSTRIAL SECTOR

In this sector, electricity can be treated as an input into the production process of a particular industry group. Total electricity demanded then would depend on the technology of the industry and the total output of its product. In our model, we have estimated the output in value added for each industry group as a function of total employment and a changing coefficient of production. At the same time the use of electricity for every unit of value added can be determined based on a coefficient that can be estimated using region-specific data. This coefficient is also affected by the price of electricity used commercially and its substitute - self-generated electric power - which can be represented by the prices of fuel oil and coal. The flow diagram for variables acting on this sector is presented as Figure 2. The aggregate growth of employment in each industry group was not determined endogenously, but fed into this model as a set of exogenous variables obtained from another study on the subject, published by the U.S. Department of Commerce, Degraff et. al (1972).

V COMMERCIAL SECTOR

In this sector, the number of customers in each of the five sub-sectors is determined by total personal income, population and income per capita in the region. The consumption per customer in each group was specified as being dependent on prices of electricity and natural gas, and a time trend specific to each sub-sector. In the case of the commercial housing sector, additionally, income per capita was specified as an independent variable, since customers of electricity in this group would be subject to considerations similar to customers in the residential sector. Figure 3 shows the factors influencing demand in this sector.

Variables used to drive the model in simulating the period up to 1990 are total personal income in the region, employment by industry groups and U.S. income per capita. This last variable was used as a determinant of net migration, on the basis that the decision to migrate from or into the region would be based on the income differential between the rest of the U.S. and this region, and on the growth in industrial employment. Birth rates were specified as being dependent on income per capita within the region, and death rates on a time trend representing the effects of greater urbanization, living patterns and advances in health care. Changes in total population were then simply determined by birth and death rates and net migration. The macroeconomic variables such as regional total personal income and national income per capita are extracted from the Commerce Department study mentioned earlier; and the endogenously determined value of population when dividing the value of regional total personal income yields income per capita for the region. Growth in employment in each industry is also derived from the same study for use in our model. Since prediction of economic variables at this time of complex national and international developments is fraught with uncertainties,

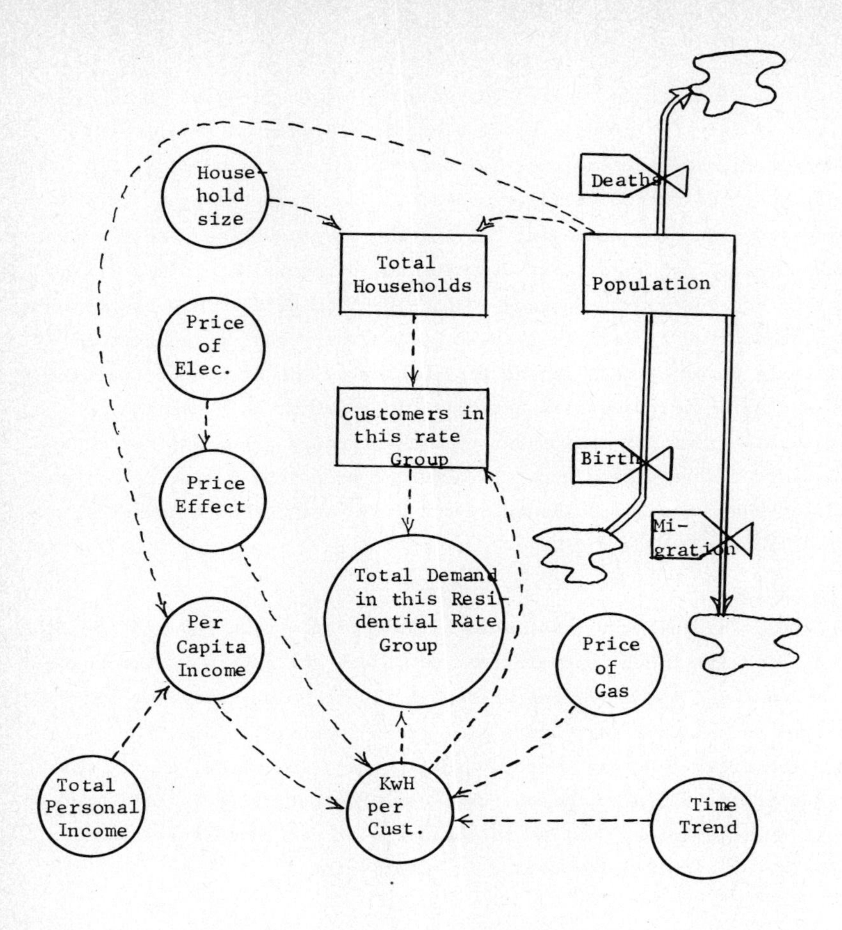

Figure I. Flow Chart for One Rate Group
In the Residential Sector

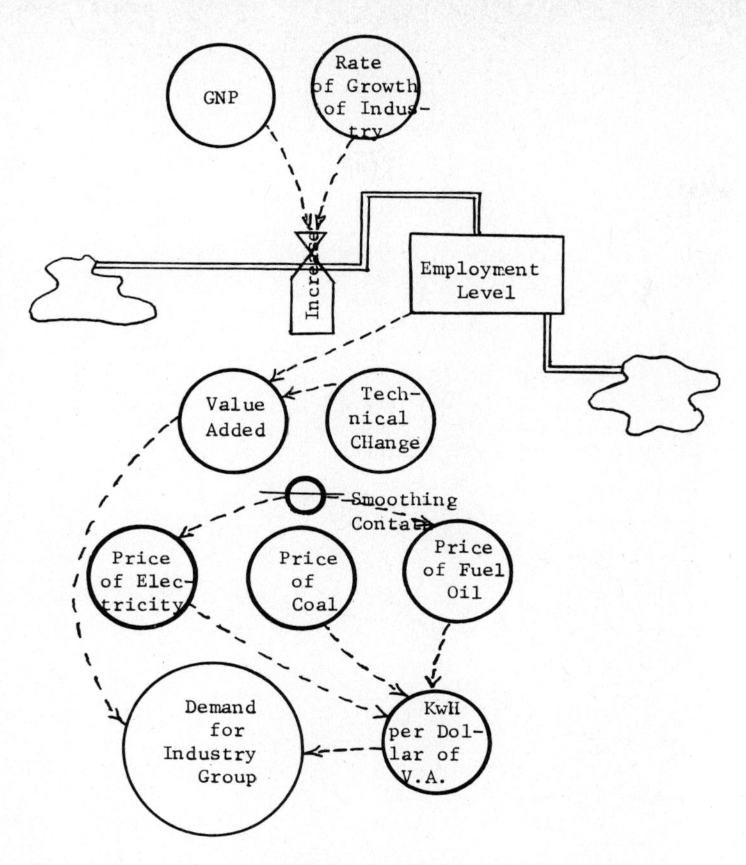

Figure II. Flow Chart for One Industry Group In the Industrial Sector

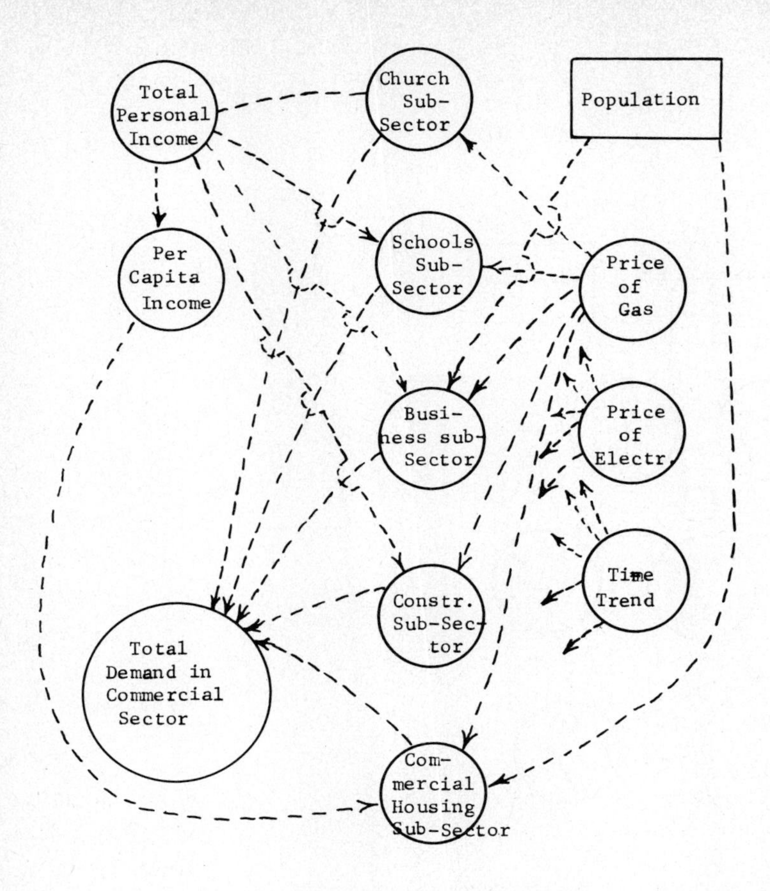

Figure III. Flow Chart for the Commercial Sector

in our simulation runs we varied the values of economic growth in order to assess the range of outcomes of key variables in response to variations in exogenous economic factors.

VI ESTIMATION OF THE MODEL

The estimation of our model was carried out using three complementary methods for different sets of relationships specified. These were (1) Ordinary least squares (OLS), (2) Simulation, and finally, (3) what may be described in Systems Dynamics simulation terminology as fine tuning of the aggregate model. The use of OLS was limited only to those relationships where the problem of simultaneity would not arise on account of any feedback effects between two equations. These equations were typically those where the independent variable represented a macroeconomic quantity, which, though a determinant of a number of separate dependent variables, would not in turn be affected by them. An example can be seen in the equation for the number of customers in the business sub-group of the commercial sector which includes total regional personal income as an independent variable. It would be reasonable to assume for the region as a whole, that the total personal income in turn is not influenced by the number of customers itself. The equations estimated by OLS are described below. These equations were adopted finally in our model since they yielded the most significant values of the coefficients estimated in comparison with other alternative forms that were specified in each case.

Most of the other equations were estimated by simulation such that predicted values for a given period in the past fitted as closely as possible with actual data observed for the same duration of time. Some values such as the lags in migration and the smoothing constant used in introducing exponentially smoothed prices into the electricity consumption equation in the industrial sector were used and adjusted by "fine tuning." Since the purpose of this study was not to estimate any elasticities of demand and because considerable research has been done in this area recently, we adopted values of elasticities estimated by Halvorsen (1973c) for the residential and commercial sectors and by Wilson (1969) for the industrial sector respectively. Some of the major equations estimated by OLS and adopted in our model are given below.

Birth Rates:

$$B_t = \underset{(.00239329)}{.02709020} - \underset{(.00000109)}{0.000000294}\, Y_t \tag{1}$$

$$R^2 = .5925$$

where: B_t = the birth rate per thousand population in period t

Y_t = income per capita for the region during period t in real dollars.

The figures shown in parenthesis in each case represent standard errors of estimates.

Migration:

$$M_t = \underset{(102371)}{291954} - \underset{(16.2418)}{130.21714}\,(Z_{t-1} - Y_{t-1}) + \underset{(343644.65)}{1316773.18}N_t \qquad (2)$$

$$R^2 = .9722$$

where:[2] M_t = net migration into/out of the region in year t

Z_{t-1} = U.S. per capita income in year t-1 in real dollars

N_t = increase in industrial employment in period t as a fraction of of employment in period t-1, and

Y_{t-1} = as defined before, lagged one year.

Business Sub-Sector Customers:

$$Bu_t = \underset{(2320.32737)}{24266.40575} + \underset{(0.55281)}{.93225}\,Y_t \qquad (3)$$

$$R^2 = .9827$$

where: Bu_t = Number of customers in business and

Y_t = as defined before

Schools and Educational Institutions Customers:

$$S_t = \underset{(227.4049)}{243.5159} + \underset{(.05418)}{.397198}\,Y_t \qquad (4)$$

$$R^2 = .9149$$

where: S_t = Number of customers in schools and educational institutions, and

Y_t = as defined above.

Church Customers:

$$C_t = \underset{(42.3145)}{4011.1165} + \underset{(.01035)}{.34975}\,Y_t \qquad (5)$$

$$R^2 = .9947$$

where: C_t = Number of customers in the church sub-sector, and Y_t as defined before

Commercial Housing Customers:

$$H_t = \underset{(3388.9335)}{-12472.4288} + \underset{(1.2408)}{6.0228}\,Y_t \qquad (6)$$

$$R^2 = .8548$$

where: H_t = Number of customers in the commercial housing sub-sector, and

Y_t as defined before

Construction Customers:

$$CO_t = -\underset{(2864.895)}{7881.649} + \underset{(1.0489)}{4.3816}\,Y_t$$

---

[2]The data set used for estimation of this equation pertained to the period 1960-69. Hence the values of per capita income applied to each of the years 1959-68.

where: CO_t = Number of customers in the construction sub-sector, and

Y_t as defined before

The effect of prices and income in this model were assumed to follow an inverted[3] - V lag structure. This is in keeping with Halvorsen's specification, although he obtained very similar estimates of elasticity with other lag structures as well, mainly on account of the high degree of autocorrelation in the price and income data used by him.

The equations shown above represent only some of those that were estimated and included in the simulation model. It is not possible in a paper of this length to cover completely the specification and estimation of a large regional model of the type dealt with herein. A similar but more comprehensive approach has been developed in exploring the future of the electrical energy sector in the U.S. in Pachauri (1975).

VII SIMULATION METHODOLOGY AND FORECASTS FOR THE REGION

The entire model referred to above was cast into a Systems Dynamics computer program using the FORTRAN WATFIV compiler and a series of sub-routines developed by Professor R.W. Llewellyn (1965). Systems Dynamics has become an increasingly popular planning and management tool in spite of a large number of applications where it has been used to represent social or economic systems without due regard to underlying economic theory and valid estimation procedures. A typical example of this type of literature is found in familiar titles such as the "Limits to Growth" and other Club of Rome Studies. The emphasis on intuitive formulation and rigidity of coefficients used in such dynamic models detracts from their closeness to reality and imparts pronounced subjectivity into their underlying assumptions and the results obtained from them. In our model we have used Systems Dynamics merely as a convenient simulation tool, estimating the various equations specified either by econometric techniques or as a result of simulation itself. Finally, the model was run with 1968 as the initial base period and simulated values for 1968-74 compared with actual observed data to validate the model. The values thus obtained were all found to be well within a 1% range of error for the time path traced for the entire 1968-73 period.

Assumptions on future growth of prices used in the study were based on a recent study by a major consulting organization in the U.S. These were modified

---

[3] An inverted V-lag structure specifies the effects of price and income changes being spread over the period of the lag in the shape of an inverted - V; that is, the effect is almost zero to start with, then gradually increases to a peak (represented by the apex of the inverted - V) and then gradually reduces to a level of zero at the end of the lag period.

after discussion with utility officials and colleagues active in the field of energy economics. We adopted a cumulative rate of growth of 5.52% per annum in real terms for electricity and coal prices and 10.64% per annum for fuel oil prices in the region. This price forecast is not defensible on any identifiable grounds and merely represents a compromise of estimate, conjecture and opinion. As such, the forecasts using this assumption were supplemented with different variations in these values to determine the effect on key output variables. With the price assumptions mentioned above, we found the increase in demand for electricity between 1974 and 1990 to be 111.10%. Most utilities in the Southeast part of the U.S.A. have been experiencing, till quite recently, growth rates in the neighborhood of 14% per annum. A 14% growth rate in the same period would result in a total increase of 185.26%. Hence, a considerable slow-down in the rate of growth in demand is predicted up to 1990.

The sensitivity runs with the same model used variations in (1) regional growth, (2) national growth and (3) growth in electricity prices, respectively. The results from these runs indicate that variations in electricity prices would have the maximum impact on demand in the residential sector, and a surprisingly small effect on the commercial sector. All three sectors are, however, sensitive to variations in the rate of regional economic growth. The model also predicted a total growth in population of 20.53% and in income per capita of 78.8% respectively over the period 1974-90. This represents an annual real growth rate of 7.5% approximately in income per capita - a prediction which appears to agree with the widespread view indicating that the Southeast part of the U.S. is currently one of the faster growing areas in the nation.

VIII TOWARDS A RATIONAL ELECTRICAL ENERGY POLICY

The model described above was extended further in the development of certain policy options, firstly, to include seasonal variations in demand that take place from month to month. A large percentage of capacity in kilowatts existing and being planned for in the electric utility industry is usually provided as a reserve for meeting seasonal peaks in demand (12-20% of average load by the utilities serving Southeast U.S.A.). The model was then run with a seasonal pricing input first with a maximum 20% differential in prices between months with extreme variations in demand. Allowing for a full 9 year inverted - V adjustment period we found that in 1990 the ratio of demand in the peak two-month period without seasonal pricing to the monthly average for the whole year was approximately 6%, but with this seasonal pricing scheme it was reduced to 0.15%. When we increased the differential in prices between the two extreme months to 25%, we found the peak demand was increased to 2.5% of the monthly average for the year. Thus, a seasonal pricing scheme must be chosen carefully such that peaking effects are not exaggerated by seasonal differences in price that are larger than those that would yield the best long run effects.

A number of simulations were also carried out with different rate structures. The concept of decreasing block rate schedules for electricity has usually been justified on the basis of decreasing marginal costs of electricity supplied to a particular customer. This concept is now being challenged in the light of higher capital costs per Kw of capacity and increases in seasonal peaks, mainly because the larger users of electricity are increasingly doing so at irregular periods, requiring the provision of extra capacity and reserves mainly to meet the demands of such customers. Thus, the actual costs of supplying extra units of power become increasingly expensive to the utility. Substantial reductions in demand were found to occur right through the simulation period, as a result of minor changes in the rate schedules for the larger users. In an industry starved for capital such effects would have a large country-wide impact in stabilizing the worsening financial picture of investor-owned public utilities.

To sum up, to a large extent the current problems of the electric utility industry in the U.S. have been brought about by a sudden increase in marginal cost of generation due to unforeseen higher prices of conventional fuels, a sharp slowdown in recent growth in demand, and increasing peaks in demand. Whereas the movement towards nuclear power is a solution to the first of these problems, this would not be without serious financial strains. Hence, a slowdown of the mushrooming capital needs of the industry is most desirable. This can be achieved by modifying present rate structures, in addition to the increase in overall electricity prices which is taking place already. The differences in seasonal demands which too are inflating capital needs and reserve capacity requirements can be reduced by suitable seasonal pricing measures. Such measures are already being tried by a few electric utilities in the nation. At any rate a substantial slowdown in demand growth is already taking place, and even with considerably reduced expansion plans we do not foresee any danger of large scale brownouts or blackouts in the U.S. up to 1990. Such a danger can only arise if the working capital needs of the industry are not met, leading to poor maintenance and operational problems. As yet, there is no evidence of any such trend in the U.S., and indications provided by our simulation runs seem to suggest that the financial problems of the electric power industry are largely of a temporary nature.

REFERENCES

Anderson, Kent P. 1972. Residential Demand for Electricity: Econometric Estimates for California and the United States, Santa Monica, Calif: The Rand Corporation, R-905, NSF, January.

Battelle Memorial Institute. 1966. A dynamic model of the economy of the Susquehanna River Basin. August 1.

Chapman, D., T. Tyrrell, and T. Mount. 1972. "Electricity Demand Growth and the Energy Crisis," Science, November 17.

Degraff, Henry L., Robert E. Graham, Jr., and Edward A. Trott, Jr. 1972. "State Projections of Income, Employment, and Population," Survey of Current Business, U.S. Department of Commerce, Washington, D.C., April.

Edison Electric Institute. 1973. Statistical Year Book of the Electric Utility Industry for 1972, New York, N.Y., November.

Fisher, Franklin M. and Carl Kaysen. 1962. A Study in Econometrics, The Demand for Electricity in the United States. North Holland Publishing Co., Amsterdam, Holland.

Forrester, Jay W. 1961. Industrial Dynamics, M.I.T. Press, Cambridge, Mass.

Forrester, Jay W. 1969. Urban Dynamics, M.I.T. Press, Cambridge, Mass.

Forrester, Jay W. 1971. World Dynamics, Wright-Allen Press, Cambridge, Mass.

Halvorsen, Robert. 1972. Residential Demand for Electricity, Cambridge, Mass., Environmental Systems Program, Harvard University.

Halvorsen, Robert. 1937a. "Long-run Residential Demand for Electricity," Discussion Paper No. 73-6, Institute for Economic Research, University of Washington, Seattle, Wash.

Halvorsen, Robert. 1973b. "Short-run Determinants of Residential Electricity Demand," Discussion Paper No. 73-10, Institute for Economic Research, University of Washington, Seattle, Wash.

Halvorsen, Robert. 1973c. "Demand for Electric Power in the United States," Discussion Paper No. 13-13, Institute for Economic Research, University of Washington, Seattle, Wash.

Llewellyn, Robert W. 1965. Fordyn, An Industrial Dynamics Simulator. Privately published, Raleigh, N.C.

National Petroleum Council. 1973. U.S. Energy Outlook, Fuels for Electricity.

Nordhaus, William D. 1973. "World dynamics: measurement without data," The Economic Journal, pp. 1156-1183. December.

Pachauri, Rajendra K. The Dynamics of Electrical Energy Supply and Demand. Praeger Publishers, New York, forthcoming, 1975.

Samuelson, Paul A. 1966. "Interactions Between the Multiplier Analysis and the Principle of Acceleration," Chapter 18, Readings in Macroeconomics, M.G. Mueller, ed., Holt, Rinehart and Winston, Inc., New York, N.Y.

Technical Advisory Committee on Load Forecasting Methodology. 1969. The Methodology of Load Forecasting, National Power Survey, Federal Power Commission, Washington, D.C.

Turvey, Ralph. 1968. Optimal Pricing and Investment in Electricity Supply. Allen and Unwin, London, U.K.

Tyrrell, T.J. 1973. Projections of Electricity Demand, Oak Ridge National Laboratory, Oak Ridge, Tenn.

Wilson, John W. 1969. Residential and Industrial Demand for Electricity: An Empirical Analysis, University Microfilms, Inc., Ann Arbor, Michigan.

PERSPECTIVES ON SHORT TERM ENERGY SHORTAGES IN THE NETHERLANDS

P. J. J. Lesuis and F. Muller

I INTRODUCTION

Energy plays an important part in nearly all economic activities. Until recently energy was only taken into consideration indirectly, by studying the environmental repercussions of regional economic growth, e.g. Muller and Pelupessy (1971), Muller (1973), and Lesuis and Muller (1974). In the present study the attention focusses on the role of energy as an input factor in production processes.

November 1973 the Arabian States revised their policies regarding to crude oil, implementing a selective boycot and an overall reduction in their production. The direct effects on the Dutch economy were estimated by using an interregional input-output model. The reduction of the crude oil supply was translated in a potential reduction of the production of the oil refining sector in the Rijnmond area, where most of the refining capacity is located, by means of a corresponding impulse on final demand as shown by Lesuis and Muller (1973). In another concept the estimated primary effects on exogenous variables in a macroeonomic model with an assumed elasticity between the level of production and the consumption of energy was introduced by the Central Planning Bureau (1974). In the meantime the changing supply conditions in the oil market caused a drastical price increase for crude oil and oil products, also affecting the prices of other sources of energy. As a result radical changes in production, consumption and the environment might be expected.

Important features of energy policy will be a reduction in the growth of energy consumption as indicated in the Energienota (1975) and an integration of energy policy with macroeconomic objectives as production, employment and the balance of payments. The energy policy will also be viewed in relation to the sectorial policy. In this regard the development of energy intensive industries can be slowed down, favouring sectors requiring less energy. This poses the problem of selective growth in the long run. The question of selective growth is also raised when in the short term the supply of energy is relatively short in relation to demand. In such situations the short term policy will be a reduction in demand, possibly forced by a distribution policy. The distinction between short term and long term is also relevant to the substitution of sources of energy, being quite small during short term actue scarcity situations. We investigated the possibilities of an optimal distribution of energy over the economic sectors in the short term case, assuming no substitution and full substitution as extremes. The relationship between energy consumption and production is stated and analyzed within the framework of an interregional input-output model, followed by a linear optimization procedure.

II THE ENERGY BALANCE

II.1 General aspects of energy consumption

Basic information on the energy consumption can be obtained from the statistics in De Nederlandse Energiehuishouding (1972). As to the units of measurement a specification either in physical volumes or in calorific terms is available. The process of extraction of primary energy sources is defined as production of energy. From an energetic point of view these primary sources are either absorbed as final products (in the input-output jargon these might be intermediate activities), or enter the energy industry which further transforms energy. Usually the transformation of energy is an energy consuming activity in itself. These losses in transformation processes are calculated as a balance between the energy production and energy consumption of the energy industry. The productiveness of the transformation process is defined as the energy output divided by the energy input, expressed in appropriate units. Within the energy industry marked differences appear, productiveness being only 34% in the generation of electricity and 82% in the oil refining process.

The general picture of the energy pattern is specified in Table 1:

Table 1. Basic balances of Energy by sector, 1972 (Tcal)

a. coal			
production	19,684	consumption energy-industry	8,065
imports	28,819	final industrial consumption	18,085
		other final consumption	4,649
		exports	21,116
		stocks	-3,412
	48,503		48,503

b. oil			
production	16,375	consumption energy industry	140,923
imports	975,129	final industrial consumption	26,112
		other final consumption	142,987
		exports	594,676
		stocks	-28,400
		bunker-exports	115,206
	991,504		991,504

Table 1.(continued)

c. gas			
production	491,829	consumption energy industry	58,467
		final industrial consumption	103,678
		other final consumption	124,709
		exports	204,263
		stocks	712
	491,829		491,829

d. electricity			
production by:		consumption energy industries	3,322
- energy industries	38,950	final industrial consumption	15,260
- final users	3,664	other final consumption	16,894
- imports	4	own production consumed by	
		final users	3,664
		transport losses	2,218
		exports	1,260
	42,618		42,618

From the table it appears that in a calorific sense the domestic consumption is provided for by domestic production for 84%. From a national point of view this makes clear that energy price increases will only gradually effect the balance of payments, assuming equal price developments in all sectors. This is mainly due to the dominant role of the natural gas production, of which more than 40% is exported.

The coal sector fully depends on imports in a short time, when domestic production will be closed down. Also the oil sector heavily relies on imports. As a producing sector the national oil industry not only serves the domestic market, but is also an important exporter to other markets in Western Europe.

The import dependency of industries generating secondary and higher forms of energy appears from their primary energy consumption. Secondary electricity generation is basically dependent on natural gas.

II.2 The input-output classification of the energy balance

The sectorial subdivision according to the economic input-output concept is based on an earlier environmental study by Muller and Pelupessy (1971). The subdivision being chosen on the basis of energy consumption is quite adequate in this case. In addition the chemical industry has been distinguished as a different sector apart from the oil industry. The energy consumption per sector is specified in Table 2.

Table 2. Consumption of energy by sector, 1972 (Tcal)

	agriculture	oil-industry	chemical industry	metal products	metallurgy	building	public utilities	housing sector	transportation	other industries	services	final consump.	Total
coal	1,761	-	1,589	148	16,370	235	4,306	-	-	3,938	-	2,600	30,947
oil	3,700	-	5,564	1,716	4,717	1,575	14,865	-	19,993	9,083	45,859	64,566	171,638
gas	2,264	611	52,661	4,086	5,850	9,386	51,875	-	-	27,009	42,000	72,709	268,451
electricity	-	-	5,080	1,786	4,129	810	-	-	800	3,455	7,485	8,609	32,154
TOTAL	7,725	611	64,894	7,736	31,066	12,006	71,046	-	20,793	43,485	95,344	148,484	503,190
proportional composition:													
coal	22.79	-	2.45	1.91	52.69	1.95	6.06	-	-	9.05	-	1.75	6.15
oil	47.89	-	8.57	22.18	15.18	13.12	20.92	-	96.15	20.89	48.10	43.48	34.11
gas	29.32	100	81.15	52.82	18.83	78.18	73.02	-	-	62.11	44.05	48.97	53.35
electricity	-	-	7.83	23.01	13.30	6.75	-	-	3.85	7.95	7.85	5.80	6.39

The input of primary energy sources in the energy industries are included only as transformation losses in the production process. These losses may be related to total inputs however. For the electricity generation this is a rather simple affair as inputs only serve to produce one output. Applying the productiveness criterion transformation losses are allocated to several kinds of inputs with as constraints total losses for the energy industry as a whole and the input structure of the energy industries.

From Table 2 differences in the sectorial energy needs can be deducted. Also the relative dependence on specific sources of energy is given, which shows the major role of natural gas in the consumption pattern. It should be noted, however, that transformation losses in the oil industry have been deleted.

II.3 A further subdivision of the energy balance: the oil industry

In this section a further refinement of the oil sector is given. One reason for this is the absence of product homogeneity as suggested by a description in calorific terms. Moreover a relevant distinction should be made between energetic and non-energetic consumption affecting substitution possibilities. Another complicating factor is that the production pattern does not necessarily coincide with the domestic demand pattern for products. With a fixed production capacity of the oil industry, this implies that short run scarcity also affects foreign trade in oil products. Using given proportions in refining one unit of crude oil (fixed product-mix), the domestic dependency on crude oil would vary from 3.125 to 68.734 million tons, in case of no substitution and no trade in products. In fact the total consumption of oil products amounted in full substitution circumstances to the equivalent of 25.71 million tons crude oil.

The consumption of oil products by sector is shown in Table 3. Contrary to the overall view, product information is only available in weight volume and not in calorific terms. In principle, however, using the right conversion factors and applying the correct subdivision in consumption, the sectorial volume consumption equals the sectorial consumption in calorific terms.

As the deliveries from the oil industry include only net production, the transformation losses in production have been deleted. Refinery gas is a by-product fully consumed in domestic uses. On the other hand the building sector is a solely user of bitumen and the chemical industry of naphta as a raw material.

II.4 Energy and input-output analysis

In the preceding sections the total sectorial energy consumption was given. Assuming a linear relationship between the sectorial production level and the sectorial energy requirement energy consumption can be related to production in the

Table 3. Domestic consumption of oil products per sector, 1972 (mln kg)*

	agriculture	oil-industry	chemical industry	metal products	metallurgy	building	public utilities	housing sector	transportation	other industries	services	final consump.	Total
refinery gas	-	-	106	-	-	-	-	-	-	-	-	-	106
LPG	-	-	77.8	7.6	2.9	1.1	-	-	0	9.7	119	218	436
naphtha	-	-	4,374	-	-	-	-	-	-	-	-	-	4,374
kerosene	-	-	-	-	-	-	-	-	10	-	-	-	10
jet fuels	-	-	0	-	-	-	-	-	793	-	-	-	793
motor petrol	21	-	-	-	-	-	-	-	230	-	1,300	1,867	3,418
special petrols/turpentine	-	-	121	-	-	-	-	-	-	-	-	-	121
paraffin	4	-	-	-	-	-	-	-	-	-	-	1,196	1,200
gasolines	300	-	68	16.6	7.4	14	50	-	1,089	178	2,565	3,038	7,326
residual oil	15	-	471	151	563	167	2,551	-	5	646	782	300	5,651
lubricants	-	-	-	-	-	-	-	-	-	-	-	211	211
bitumen	-	-	-	-	-	817	-	-	-	-	-	-	817
other products	-	-	2.8	10.4	2.1	2.9	-	-	-	21.2	-	310	350

* Sources: De Nederlands Energiehuishouding, 1972; E.E.C. Energy Statistic, Yearbook, 1960-1971.

same way as in traditional input-output analysis. Energy coefficients have been defined (either in calories or volumes) in relation to 1972 production levels (1965 prices). If E is a matrix of sectorial energy coefficients, total energy consumption for each product can be given as

$$e = E\underline{x} + \underline{fe}$$

where $\underline{e}$ is a vector of oil products, $\underline{x}$ a vector of production levels and $\underline{fe}$ a vector containing final energy consumption. From the input-output analysis it follows that

$$\underline{e} = E(I-A)^{-1}\ \underline{f} + \underline{fe}$$

$\underline{f}$ being a final demand vector and (I-A) the Leontief matrix.

Thus $E(I-A)^{-1}$ produces the energy consumption coefficients in terms of final demand, which allows the calculation of energy consumption per unit of final demand. This could be subdivided in the direct energy consumption of the ith sector to produce one unit of final demand and indirect consumption, which is the energy content of the necessary intermediate products. This is shown in Table 4.

Table 4. Sectorial energy consumption per unit of final demand, 1972. (Tcal per million Dfl. 1965)

	Total	Direct	Indirect
Rijnmond:			
agriculture	1.471012	0.625983	0.845029
oil industry	0.637998	0.	0.637998
chemical industry	6.242012	4.857551	1.384461
metal products	1.471735	0.511581	0.960154
metallurgy	2.816623	1.823052	0.993571
building,construction	1.764356	0.705888	1.058467
public utilities	17.130884	15.643004	1.487880
housing sector	0.388545	0.	0.388545
transportation	2.366913	1.748900	0.618013
other industries	2.109369	0.934434	1.174935
service sector	3.528791	2.666649	0.862142
Netherlands, rest of the country:			
agriculture	1.790545	0.625983	1.164562
chemical industry	6.286655	4.857551	1.429104
metal products	1.510208	0.511581	0.998627
metallurgy	3.137913	1.823052	1.314861
building, construct.	1.770182	0.705888	1.064294

(continued on next page)

	Total	Direct	Indirect
Netherlands (continued)			
public utilities	16.878127	15.643004	1.235123
housing sector	0.392753	0.	0.392753
transportation	2.604663	1.748900	0.855763
other industries	2.089575	0.934434	1.155141
service sector	3.531805	2.666649	0.865156

III AN INTERREGIONAL INPUT-OUTPUT ENERGY MODEL

III.1 Model description

The model enables the analysis of short term effects from a reduction of energy supply on the Dutch economy. For this purpose the technique of linear programming has been applied to an interregional input-output model. Regionally a distinction was made between the Rijnmond area on one side and the Netherlands excluding this area at the other. It may be assumed that the production of the oil refinery sector is fully concentrated in the Rijnmond area. The other sectors appear in both regions however. Apart from the usual input-output relationships a set of energy restrictions has been added. Energy use on a regional basis not being available, national coefficients have been assumed at the regional level as well. Energy shortages might be alleviated by extra imports of energy products over original use. At first labour market considerations were included as restrictions, specifying the objective function in income terms. As an alternative the objective function was specified in labour terms. The general structure of the model thus appears as follows:

$$\underset{1x1}{\text{Max } Y} = \left[\underset{1x21}{\underline{v}'} \;\; ; \;\; \underset{1x13}{\underline{o}'} \right] \underset{34x1}{\underline{x}}$$

Subject to:

$$\begin{bmatrix} \underset{21\text{x}21}{-I+A} & \underset{21\text{x}13}{0} \\ \underset{13\text{x}21}{E} & \underset{13\text{x}13}{-I} \\ \underset{21\text{x}21}{I} & \underset{21\text{x}13}{0} \\ \underset{1\text{x}21}{\underline{w}'} & \underset{11\text{x}13}{\underline{o}'} \\ \underset{1\text{x}21}{\underline{o}'} & \underset{1\text{x}13}{\underline{i}'} \end{bmatrix} \begin{bmatrix} \underset{34\text{x}1}{\underline{x}} \end{bmatrix} \begin{matrix} \leq \\ \leq \\ \leq \\ \leq \\ \leq \end{matrix} \begin{bmatrix} \underset{21\text{x}1}{-\underline{f}} \\ \bar{\underline{e}} \\ \underset{21\text{x}1}{\bar{\underline{x}}_{72}} \\ \underset{1\text{x}1}{\bar{W}} \\ \underset{1\text{x}1}{\bar{B}} \end{bmatrix} \begin{matrix} (1) \\ (2) \\ (3) \\ (4) \\ (5) \end{matrix}$$

and the non-negativity constraints $\underset{34\text{x}1}{\underline{x}} \geq \underset{34\text{x}1}{\underline{o}}$

where Y is national product;

$\underline{v}$ is a vector of value added coefficients;

$\underline{f}$ is a final demand vector;

$\underline{x}$ is a vector of decision variables, i.e. sectorial production levels and the value of imported oil products;

$-I+A$ specifies the input-output relationships;

E is a matrix of energy coefficients;

$\bar{\underline{e}}$ is a vector of energy products available;

$\bar{\underline{x}}_{72}$ is a vector of maximum allowable production levels;

$\underline{w}$ is a vector of labour coefficients;

$\bar{W}$ is labour supply available;

$\bar{B}$ is the total amount available for additional imports of energy products.

In addition imports and exports can be quantified, using sectorial import and export coefficients, as a fraction of sectorial production and final demand respectively. Apart from changes in imports serving final demand balance of payment effects are indicated in this way.

The input-output restrictions primarily serve to guarantee the production of the necessary intermediate products; on the public utility sector a fixed level of final demand has been imposed. Because of the short run character of the analysis production capacity and labour supply have been fixed at the 1972 level as maximum restrictions in (3) and (4). The second set (2) contains the energy restrictions. The available amount of energy in $\underline{e}$ depends on factors as domestic production and international trade of energy products as well as on the possibility of reductions in the final use of energy consumption. The availability of energy products is further influenced by the amount made available for additional imports in (5); it is assumed throughout the study that oil products are imported at existing 1972 export prices.

Energy shortages may arise out of a physical reduction in crude oil supply. The immediate consequence is that the quantity available for domestic refining declines in accordance with the overall supply reduction. Thus the production possibilities of the refinery sector in the Rijnmond region are limited propotionally. The influence on the other sectors depends on the assumptions made about the availability of oil products ($\bar{\underline{e}}$) and substitutions possibilities. Contrary to a direct relationship between the availability of oil products and production levels full substitution reduces the E matrix to a vector of total energy consumption coefficients. The latter are based on the calorific consumption as specified in Table II. It should further be noted that energy substitution implies a different relationship between the oil sector and the sectors receiving oil products. This implies a correction in the output row of the refinery sector depending on the share of oil consumption in total energy requirements and the reduction in the deliveries obtained from the oil sector. In the same way the value added coefficients increase due to the increased oil productivity.

III.2 Applications of the model

The alternatives investigated with the model may be summarized as follows:

	(1)	(2)	(3)	(4)	(5)	(6)	(7)
Reduction in crude supply	0%	20%	20%	20%	20%	20%	20%
Allowed amount for additional imports of oil products	0	0	0	0	ƒ100 mln	ƒ1000 mln	ƒ1000 mln
Reduction in the export of oil products	0%	20%	20%	20%	20%	0%	0%
Economizing electricity consumption in final demand	0%	20%	20%	20%	20%	20%	0%
Economizing final consumption of oil products	0%	20%	20%	20%	20%	20%	0%
Substitution	no	no	no	yes	no	no	no
Maximize value added	x	x	-	x	x	x	x
Maximize labour	-	-	x	-	-	-	-

In all cases the original imports of oil products were maintained. It seems that most rigid assumptions are made in alternatives (2) and (3), as shortages can only be met by decreasing domestic use forces by a reduction in production levels. The effects are moderated in the case of full substitution in alternative (4) or by making available additional imports of the most scarce products in (5), (6) and (7). Results of the model under different assumptions are discussed in more detail in the following section.

III.3 Model results.

Sectorial production levels, total production, imports, exports, income and employment are summarized in Table 5. A comparison with 1972 figures is made in Table 6.

Table 5. Overall model results

production per sector (mld, 1972)	(1)	(2)	(3)	(4)	(5)	(6)	(7)
Rijnmond:							
agriculture	0.236	0.236	0.236	0.236	0.236	0.236	0.236
oil industry	6.382	5.105	5.105	5.105	5.105	5.105	5.105
chemical industry	3.047	0.341	0.318	1.681	2.135	3.047	1.742
metal products	2.156	2.156	2.156	2.156	2.156	2.156	2.156
metallurgy	0.663	0.663	0.663	0.663	0.663	0.663	0.663
building, construction	2.083	0.280	0.275	2.083	2.083	2.083	2.083
public utilities	0.445	0.317	0.316	0.368	0.372	0.381	0.422
housing sector	0.477	0.477	0.477	0.477	0.477	0.477	0.477
transportation	4.561	4.561	0.482	4.561	4.561	4.561	4.561
other industries	4.119	4.119	4.119	4.119	4.119	4.119	4.119
service sector	6.376	1.598	3.192	6.376	6.376	6.376	6.376
Netherlands, rest of the country:							
agriculture	13.788	13.788	13.788	13.788	13.788	13.788	13.788
chemical industry	9.217	8.811	7.325	9.217	9.217	9.217	9.217
metal products	15.278	15.278	15.278	15.278	15.278	15.278	15.278
metallurgy	17.240	17.240	17.240	17.240	17.240	17.240	17.240
building, construction	19.970	18.405	18.409	19.970	19.970	19.970	19.970
public utilities	3.913	3.298	3.268	3.371	3.373	3.376	3.905
housing sector	4.151	4.151	4.151	4.151	4.151	4.151	4.151
transportation	9.438	7.044	3.472	9.438	9.438	9.438	9.438
other industries	48.284	48.284	48.284	48.284	48.284	48.284	48.284
service sector	42.220	39.115	42.220	42.220	42.220	42.220	42.220
Total production	214.054	195.018	190.785	210.793	211.252	212.177	211.442
Total imports	31.600	28.890	27.434	30.639	30.878	32.059	31.700
Total exports	49.163	45.093	40.253	47.777	48.047	49.309	48.488
Total income	104.038	93.627	92.258	103.303	102.865	103.207	103.100
Total labour needs (millions)	4.221	3.826	3.850	4.207	4.209	4.213	4.214

Table 6. Reduction in production levels and changes in some macro-economic variables.

Production decline (mld, 1972)	(2)	(3)	(4)	(5)	(6)	(7)
Rijmond:						
oil industry	1.277[1]	1.277[1]	1.277[1]	1.277[1]	1.277[1]	1.277[1]
chemical industry	2.706	2.729	1.366	0.912	-	1.305
building, construction	1.803	1.808	-	-	-	-
public utilities	0.128[1]	0.129[1]	0.080[1]	0.077[1]	0.064[1]	0.023
transport	-	4.079	-	-	-	-
other industries	0.260	-	-	-	-	-
service sector	4.778	3.184	-	-	-	-
Netherlands, rest of the country:						
chemical industry	0.406	1.892	-	-	-	-
building, construction	1.565	1.560	-	-	-	-
public utilities	0.615[1]	0.615[1]	0.542[1]	0.539[1]	0.536[1]	0.008
transport	2.394	5.966	-	-	-	-
service sector	3.105	-	-	-	-	-
change in production	-19.036	-23.269	- 3.261	- 2.802	- 1.877	- 2.612
change in balance of payments	- 1.360	- 4.744	- 0.425	- 0.394	- 0.263	- 0.775
change in income	-10.411	-11.780	- 9.735	- 1.173	- 0.831	- 0.938
change in labour needs (millions)	- 0.395	- 0.371	- 0.014	- 0.012	- 0.008	- 0.007

[1] By assumption

Also apart from the reduction in the oil sector consequences are severe when there are no opportunities for substitution or additional imports (alternatives 2 and 3). The building industry is badly hit in both alternatives by lack of bitumen as a raw material. In comparison with employment, maximizing the national product favours transportation but harms the service sector. At the aggregate level both alternatives show a considerable loss in income and labour demanded. Employment maximization leaves 24,000 additional working places untouched, but the losses in income (1.365 milliard) and the balance of payments (3.384 milliards) are exorbitant. In both cases (2) and (3) production is limited by lack of motor petrol, residual oil and bitumen, and in the case of value added maximization also jet fuels.

Scarcity and affluence exist at the same time: while some oil products do limit production possibilities, not all oil products are exhausted. This indicates that substitution might be fruitful. To examine this situation full substitution as assumed in alternative (4). In this alternative the effects mainly appear in the oil industry and the public utility sector as a consequence of limitations in the final demand of electricity. The only other sector where production is affected is the chemical industry in the Rijnmond area.

Also an extra amount of money set available for the importation of extra oil products relieves considerably, as shown in alternatives (5), (6) and (7). In all these cases the loss in employment, national product and the balance of payments is moderate. In run (6) maintenance of the original imports of oil products together with the possibility of extra imports to the amount of Dfl. 1 milliard and a reduction in domestic fuel consumption limits the production losses only to the public utility sector and the oil industry.

Together with no domestic reductions in fuel energy consumption in alternative (7) the main effect is a decline in the chemical industry in Rijnmond, affecting total production and the balance of payments.

It is also interesting to see how the amount available for additional imports is allocated to various products. This allocation has been specified in the following table:

Table 7. Additional imports of oil products (Dfl. mrd)

Product:	(5)	(6)	(7)
refinery gas	-	-	-
LPG	2.387	12.895	16.458
naphta	3.625	120.847	74.415
kerosene	0.138	4.229	4.367
jet fuels	13.463	63.350	61.798
petrol	24.843	67.482	105.038
special petrol, turpentine	0.405	6.771	5.417
paraffin	-	2.736	28.008
gasolines	29.744	291.298	353.220
residual oil	14.566	22.805	240.678
lubricants	-	21.939	39.422
bitumen	10.589	16.342	16.545
other products	0.235	169.300	54.626
Available amount (Dfl, millions)	100	1000	1000
$\lambda = {}^{\partial Y}/\, \partial B$	0.0094	0.0000	0.0094

From Table 7 it is clear that several products are imported only to a small degree (e.g. kerosene, special petrols). Considerable amounts are spent on gasolines, residual oil, petrol and naphta's. In all cases the total amount set available has entirely been used. The division between the different kinds of use secures the same marginal return of each additional amount available for each products equal to $\partial Y/\partial B$. The effect of supplying an additionaly amount reduces sharply when scarcity in energy products reduces, starting with a shadow price of 2.699 in alternative (2) and becoming zero ultimately.

IV CONCLUSION

Several different outcomes of a reduction in crude oil supply on the Dutch economy may occur. When no additional imports of oil products are available and no substitution opportunities exist the consequences of a reduction in oil supply for production, employment and the balance of payments are severe. The only possibilities for improvement in this case are economizing and optimization. An optimization criterium in terms of employment rather than value added gives an additional decrease in income and a further deterioration of the balance of payments; the gain of extra employment is relatively small.

When domestic reductions in the final consumption of oil products can be realized in combination with additional imports of oil products only a small amount of money is needed to alleviate these effects considerably. The same result is obtained when a high degree of substitution can be realized, which may be very difficult however, especially when oil products are used as a raw material. The shadow prices of the oil products indicate where further savings in domestic use, extra importation of oil products and substitution may be worthwhile.

In order to achieve the necessary reductions in final energy use and to bring about an optimal allocation of oil products a distribution policy might be necessary.

REFERENCES

Central Bureau of Statistics, 1972; De Nederlandse Energie Huishouding, Staatsuitgeverij, Den Haag, 1972.

Central Planning Bureau, 1974; Nota over de beperking van de olie-aanvoer en de gevolgen daarvan (unpublished paper).

Lesuis, P.J.J. and Muller, F., 1973; De economische gevolgen van de oliecrisis, Economisch-Statistische Berichten, vol. 58, pp. 1028-30.

Lesuis, P.J.J and Muller, F., 1974; Predicting air pollution levels in the Rijnmond area; experience with a multiple source dispersion model, Paper prepared for the WHO-working group on regional residuals environmental quality management modelling, Rotterdam.

Ministry of Economic Affairs, 1975; Energienota; kamerstuk 13.122 nrs. 1 en 2 zitting 1974/1975, Staatsuitgeverij, Den Haag.

Muller, F., 1973; An operational mathematical programming model for the planning of economic activities in relation to the environment, Socio-Economic Planning Sciences (The Pergamon Press), vol. 7, pp. 123-138.

Muller, F. and Pelupessy, W., 1971; Economische waardering van de schaarse lucht in Rijnmond, Economisch-Statistische Berichten, vol. 56, pp. 293-306.

Statistical Office of the European Community, 1973; Energy statistics, no. 1-2.

Statistical Office of the European Community, 1971; Energy statistics, Yearbook, 1960-1971.

A LINEAR PROGRAMMING MODEL FOR DETERMINING AN OPTIMAL REGIONAL DISTRIBUTION OF PETROLEUM PRODUCTS

Craig L. Moore and Andris A. Zoltners

I INTRODUCTION

The United States has always enjoyed a relative abundance of national resources. Except for brief periods,usually associated with a war, there has never been a prolonged shortage of a basic commodity in America. During the few periods where shortages have occurred, the government has usually turned to price controls and rationing rather than relying on the market mechanism to allocate these resources.

Since 1972 there has been a growing concern over the burgeoning shortage of petroleum. Production and refining capacity has not increased at a sufficient rate to meet swelling demands, and the 1973 Middle East war and Arab oil embargo against the industrialized West compounded this problem. In the long run, new sources of energy or improved technology may correct this situation, but many countries will face a continuing shortage of petroleum products, high prices and constraints on development for at least the next several years.

In 1973, the American government instituted price controls and established a mandatory allocation program. This allocation plan was primarily voluntary and dealt only with broad aggregate categories of petroleum users. The formulation of a detailed distribution program, which would include both industrial and regional allocations, requires much more information than is presently available and the development of sophisticated tools for economic planning and analysis.

This paper examines the potential use of linear programming in developing an oil allocation plan. The primary objective is to provide an economic model and a method of analysis which will help planners and policy makers determine the most efficient distribution of petroleum products among industries and regions in a nation.

A second objective is to develop an economic model for simulating the impact of a particular allocation program on the economy at both the national and regional levels. This is highly desirable because of the critical impact such programs have on prices, employment, production, and the basic quality of living.

While this general approach can be applied to almost any scarce resource, this paper will be limited to the petroleum allocation problem. This is done for two reasons. First, the distribution of petroleum products is a problem that will likely arise again. Petroleum serves as the backbone of energy production in most industrialized nations and, therefore, it is of critical importance to have the capability of formulating an allocation plan which will both optimize growth and maintain economic stability. Second, it is easier for most people to relate a model to a concrete problem rather than to deal with it entirely in abstract terms. The allocation of petroleum, therefore, serves as an illustration of a more general technique which could be applied to agricultural products, land, metals, water, or any other resource which is vital to society.

II A GENERAL DESCRIPTION OF THE MODEL AND ITS ASSUMPTIONS

The distribution process described in this paper distinguishes between three types of uses. The first category includes special allocations for the production of goods and services with a high social priority such as heating oil for hospitals, gasoline for police cars, or petrochemical feedstocks for manufacturing drugs. Home heating oil could be included in this category because it is a necessity for millions of American families and the use of a price mechanism for its allocation would be regressive and work a hardship on low income families. Allocations for these priority uses are made off the top of any existing supply of petroleum products and therefore are not included in the model.

Second, there are many goods and services (both private and public) which society must maintain on a minimum level. These would include final consumption goods such as clothing, agricultural products, consumer durables, and certain governmental services, as well as the intermediate goods necessary for their production. In addition, a minimum level of production of many basic goods is essential to minimize inflationary pressures arising from shortages. Petroleum allocations for the production of all of these goods and services are insured through the incorporation of a set of constraints in the model.

The third category of petroleum products includes discretionary uses. All those goods and services not included in the first two groups would fall into this category. The model assumes that there are n groups or industries vying for this discretionary amount. These groups are based on the Standard Industrial Classification Code of the U.S. Department of Commerce augmented by some additional special categories. They will be indexed by the letter i, i.e. $i = 1,2,\dots,n$. To simplify the description of the model, each group will hereafter be referred to as industry i regardless of the actual economic function.

Since the model allocates spatially, users are further divided by region.

Each industry has a different regional concentration of productive capacity. This not only causes variation in the regional demand for petroleum products, but can also create differences in the impact on employment between regions during a period of reduced output.

Regions will be indexed by the letter j, i.e. $j = 1,2,\ldots,m$, where the region partition specifies m regions in the nation. A logical division may be by state, which would mean that m would equal 51 if we include the District of Columbia.

The model assumes either a predetermined level of available crude oil or predetermined levels of the four major refined petroleum products: fuel oil (also called distillates, heating oil), gasoline, heavy fuel oil (also called bunker fuel, residual fuel), and petrochemical feedstocks. If a predetermined level of crude is assumed, then a model is presented for determining levels for the various petroleum products by establishing their market prices so as to equate refinery supply with consumer demand. This model is a nonlinear programming model which incorporates an aggregate refinery LP and demand equations and will be discussed later in the paper.

The four major refined petroleum products will be indexed by the letter k, i.e. $k = 1,2,3,4$. Classes $k=1$, $k=2$, $k=3$, $k=4$ are fuel oil, gasoline, heavy fuel oil, and petrochemical feedstocks, respectively.

The LP model is a single period model. Typically, the demand for petroleum products and output levels for goods and services will be specified for a planning horizon which is flexible (it may be a week, month, quarter, year, etc.).[1]

The decision variable in the model is output. Output is related to employment by a set of linear employment functions. The sum of these functions serves as the LP objective function. There are five linear constraint categories in the model. First, industry input-output relationships insure a balanced economy. Minimum production levels are maintained so that the economy does not incur shortages of vital goods and to discourage supply based inflation. Another set of constraints assures that the available supply of refined petroleum products is not exceeded. Industry capacity constraints are included. Finally, regional economic constraints make sure that regional economies remain healthy. A statement of the model follows.

---

[1]Multi-period extensions of the model, incorporating inventory policies and lag effects, remains the subject for future research.

III THE MODEL

$$\text{Maximize} \quad \sum_{i=1}^{n} \sum_{j=1}^{m} M_j e_{ij} O_{ij} \tag{1}$$

subject to:

$$O_i - \sum_{j=1}^{m} O_j = 0 \quad \text{for } i = 1,2,\ldots,n \tag{2}$$

$$O_i - I_i^+ + I_i^- - EXP_i + IMP_i \geq DEMAND_i \quad \text{for } i=1,2,\ldots,n \tag{3}$$

$$O_i \geq \bar{O}_i \quad \text{for } i = 1,2,\ldots,n \tag{4}$$

$$O_i - \sum_{h=1}^{n} \beta_{ih} (O_h - I_h^+ + I_h^- - EXP_h + IMP_h - DEMAND_h) = 0 \quad \text{for } i = 1,2,\ldots,n \tag{5}$$

$$I_i^+ - I_i^- \leq \bar{I}_i \quad \text{for } i = 1,2,\ldots,n \tag{6}$$

$$I_i^- - I_i^+ \leq \tilde{I}_i \quad \text{for } i = 1,2,\ldots,n \tag{7}$$

$$EXP_i \geq L_i \quad \text{for } i = 1,2,\ldots,n \tag{8}$$

$$IMP_i \leq U_i \quad \text{for } i = 1,2,\ldots,n \tag{9}$$

$$\sum_{i=1}^{n} \sum_{j=1}^{m} r_{ijk} O_{ij} \leq S_k \quad \text{for } k = 1,1,3,4 \tag{10}$$

$$O_{ij} \leq C_{ij} \quad \text{for } i = 1,2,\ldots,n; \text{ for } j = 1,2,\ldots,m \tag{11}$$

$$O_{ij} \geq M_{ij} \quad \text{for i and j where region j has a high dependence on industry i} \tag{12}$$

$$\sum_{i=1}^{n} e_{ij} O_{ij} \geq \frac{E_j + L_j (M_j - 1)}{M_j} \quad \text{for } j = 1,2,\ldots,m \tag{13}$$

$$I_i^+, I_i^-, IMP_i \geq 0 \quad \text{for } i = 1,2,\ldots,n \tag{14}$$

where:

O_{ij} is the total output level for industry i in region j the planning period. O_{ij} is the decision variable.

O_i is the total output of industry i in the planning period as defined in (2).

e_{ij} is the employment/output ratio for industry i in region j. (This ratio and several others used in the model will be discussed in detail later.)

M_j is the employment multiplier for region j.

β_{ij} is the fraction of output in industry h that is consumed by industry i as a production input.

Note that $0 \leq {}_{ij} \leq 1$ and $\sum_{i=1}^{n} \beta_{ih} = 1$.

I_i^+ is the inventory accumulation in industry i in the planning period.

I_i^- is the inventory depletion in industry i in the planning period.

EXP_i is the amount of goods produced in industry i which are exported in the planning period.

IMP_i is the amount of goods produced for industry i which are imported in the planning period.

$DEMAND_i$ is an estimate for the demand for the goods produced in industry i for government purchases, households, gross private capital formation, etc. (demand sector).

$\bar{O}_i$ is the minimal level of output required for industry i for the planning period.

$\bar{I}_i$ is an upper bound for net inventory change for industry i.

$\tilde{I}$ is a lower bound for net inventory change for industry i.

L_i is a lower bound for exports for industry i.

U_i is an upper bound for imports for industry i.

r_{ijk} is the consumption/output ratio for industry i in region j for fuel oil (k=1), gasoline (k=2), heavy fuel oil (k=3), and petrochemical feedstocks (k=4).

S_k is the total supply of fuel oil (k=1), gasoline (k=2), heavy fuel oil (k=3), and petrochemical feedstocks (k=4) available in the planning period.

C_{ij} is the output capacity for industry i in region j.

M_{ij} is the minimum level of output for industry i in region j for the planning period, where region j has a high dependence on industry i.

E_j is the minimal level of employment required in region j for the planning period.

L_j is the labor force in region j.

The model maximizes employment. The objective function (1) is the sum of employment functions, $e_{ij}O_{ij}$, adjusted by a regional employment multiplier M_j, designed to reflect secondary and tertiary employment effects.[2] The employment functions relate the leve of output, O_{ij}, to employment in industry i in region j. For practical purposes, due to the available data, these functions, can be assumed linear.[3] Actual estimation of the employment functions, as well as estimation of other relationships used in the model, is deferred to the next section.

There are several reasons why the objective of the model is to maximize employment. First, the flow of household income is the mainspring of our national economy. If consumer demand is weakened, our economy could face a serious recession as well as a critical shortage of petroleum. There would be high social costs associated with widespread unemployment such as welfare payments and increased crime. At the same time, revenues from income taxes would fall and the government would face serious economic problems.

---

[2]The employment multiplier has the following effect. Suppose region j has a labor force L_j=100, employment multiplier M_j=1.5 and a 10% initial unemployment due to an energy shortage. After secondary and tertiary effects are accounted for, the ensuring unemployment rate would be 15% (10% x 1.5).

If employment is to be maximized the following objective function is required:

$$\text{maximize} \quad \sum_{j=1}^{m} L_j - \sum_{j=1}^{m} M_j (L_j - \sum_{i=1}^{n} e_{ij} O_{ij})$$

or equivalently,

$$\text{maximize} \quad \sum_{j=1}^{m} L_j (1-M_j) + \sum_{i=1}^{n} \sum_{j=1}^{m} M_j e_{ij} O_{ij}$$

Since $\sum_{j=1}^{m} L_j (1-M_j)$ is a constant and hence unaffected by the optimization, the objective function (1) becomes

$$\text{maximize} \quad \sum_{i=1}^{n} \sum_{j=1}^{m} M_j e_{ij} O_{ij}$$

[3]If non-linear functions, $f_{ij}(O_{ij})$, are deemed more suitable or even necessary, the model becomes a non-linear programming model. The simplest of these models would be a separable programming model in which f_{ij} is assumed concave and piecewise linear.

Additional complexities arise if whole industries are allowed to shut down in some region (i.e. constraints (17) and (13) removed). An integer programming model becomes appropriate to portray the jump in employment when output becomes positive. In this case, if f_{ij} is approximated by a piecewise linear function, a separable programming model can be used.

Constraints (3) stipulate that each industry provide enough output to meet demand, maintain an adequate inventory and maintain a reasonable import-export position. Since government has some control over the demand section via monetary and fiscal policy, demands can be established which prevent shortages of key commodities, retard recessionary pressures, and control inflation.

Constraints (4) stipulate that for each industry, minimal levels of output must be obtained. These minimums are set at the national level to prevent shortages of key commodities, retard recessionary pressures and to control inflation.

One possible strategy for determining these minimums would be to start with a list of all final goods and services and determine at what point a decrease in their supply would cause an undesirable rise in price. With these levels determined, a national input-output matrix could be used to estimate the production of all intermediate goods that would be required to provide the desired levels of final goods. This would result in a set of constraints (4) for each industry to be met by the model.

Constraints (5) are the input-output equations necessary to strike a balance between final and intermediate goods at the national level. Without these constraints, the maximization of employment would be the only criteria for allocation and petroleum supplies would be channeled into an industry regardless of the demand for the output. Without this constraint, therefore, the model could reach a solution having serious imbalances in production with shortages of some goods and excess inventories of others. Implementation of such a solution could lead to serious unemployment and the model's objectives would be negated.

Constraints (6) and (7) restrict the allowable net inventory change. The national inventories for industry i are allowed to increase at most $\bar{I}_i$ and decrease by no more than $\tilde{I}_i$. This is an important consideration if policy makers determine some margins of safety which should be maintained for key goods.

Constraints (8) and (9) are trade restrictions. Imports for industry i are restricted to be no more the U_i and the production of exports is constrained to be at least L_i. These constraints are added to insure that the nation maintains a reasonable balance of payments position. Such constraints would be set as a matter of economic policy.

Four consumption functions, $r_{ijk}O_{ij}$, relate the average amount of fuel oil (k=1), gasoline (k=2), heavy fuel oil (k=3), and petrochemical feedstocks[4] (k=4) required per unit of output in industry i in region j. These are assumed to be linear relationships. Constraints (10) reflect that available supplies, S_k (k=1,2,3,4), of fuel oil, gasoline, heavy fuel oil, and petro-chemical feedstocks can not be exceeded.[5] The coefficient r_{ijk} is the consumption coefficient for petroleum product k in region j and industry i. Within an industry this ratio primarily reflects climatic, technological, and scale variations between regions. For example, if region 1 is warmer than region 2 during the same planning period, the r_{i11} will be less than r_{i21}. If industry i in region j does not use petroleum product k, then $r_{ijk} = 0$.

Constraints (11) are production capacity constraints in industry i in region j. While overall production in the aggregate is not expected to rise when the supply of petroleum is reduced, it is likely that some shifts in production between industries and regions will be desirable from the standpoint of efficiency. Without some constraints on capacity, the model would shift all of the production within an industry to the most efficient region. Consequently, constraints (11) insure that no more production is shifted into a region than it has capacity to produce. These constraints are based on estimates of past output made by the Department of Commerce by region and S.I.C. category.

---

[4] There are many petrochemical feedstocks, e.g. ethane, ethylene, propylene, benzene, toluene, zylene, etc. To simplify the model, the petrochemical feedstocks have been aggregated. This can be done since refineries have the capability to swing from any one of these products to any other on roughly a one-to-one basis [7]. If necessary, however, petrochemical feedstocks can be disaggregated within the framework of a more complex linear programming model.

[5] It should be noted that constraints (10) assume complementarity between fuel oil and heavy fuel oil. They do not assume substitutability. If one can be substituted for the other and vice-versa, then output O_{ij} has to be decomposed into O_{ij1} and O_{ij3} where O_{ij1} is that portion of output consequent on the use fuel oil and O_{ij3} is that portion related to heavy fuel oil. Then constraints (10) should be replaced by:

$$\sum_{i=1}^{n} \sum_{j=1}^{m} r_{ijk} O_{ijk} \leq S_k \qquad \text{for } k = 1, 3$$

$$\sum_{i=1}^{n} \sum_{j=1}^{m} r_{ijk} O_{ij} \leq S_k \qquad \text{for } k = 2, 4$$

$$O_{ij} + O_{ij1} + O_{ij3} \qquad \text{for } i = 1,2,\ldots,n; \; j = 1,2,\ldots,m$$

If fuel oil and heavy fuel oil are substitutable and it is desirable to have one or the other, but not both, the model becomes an integer programming model.

Constraints (12) and (13) assure that each region maintains some subsistence level of economic activity. Constraint (12) sets a lower limit on output in key regoinal industries. This would be useful in protecting regional economic stability where the export base of the region was very dependent upon one or a few industries. Location quotients or an index of industrial concentration could be used in identifying such circumstances [2].

Constraints (13) set a lower bound E_j on the employment of each region. This bound can be determined by applying a standard maximum unemployment rate, μ, to each region's labor force, L_j:

$$E_j = L_j\,(1-\mu).$$

The standard for maximum regional unemployment should be determined as a matter of policy. Constraints (13) are derived analagous to the objective function:

$$L_j - M_j(L_j - \sum_i e_{ij} O_{ij}) \geq E_j$$

is equivalent to

$$\sum_i e_{ij} O_{ij} \geq \frac{E_j + L_j(M_j - 1)}{M_j}$$

The regional employment multiplier, M_j, can be estimated by the size and economic composition of the region. The magnitude of the multiplier is inversely related to the proportion of the local economy which is involved in producing regional exports.[6] Failure to include the multiplier for each region could result in a serious understatement of the lower employment bound. Initial unemployment, if allowed to fall to an unadjusted minimum, would only account for initial layoffs. The reduction in income would lead, in turn, to further rounds of unemployment due to a fall in regional income.

The LP decision variable, output O_{ij}, can be viewed as the common denominator in the model. Both employment and petroleum product consumption are expressed in terms of output. In addition, regional economic constraints and input-output balance equations are set forth in terms of output. Once the optimal output levels have been resolved by the linear program, the associated allocation scheme and employment levels can be determined.

Variables, O_i (i=1,2,...,n), defined in constraints (2) are non-essential.

---

[6] M_j, the regional multiplier, equals 1/(1-kb) where kb is the propensity of regional residents to consume locally produced goods and services. kb can be estimated by the proportion of employment in the service sector of the region using a minimum requirements technique [2],[3].

They are merely a convenience for expository purposes. When actually implemented, constraints (2) would be omitted and constraints (3), (4), and (5) would be altered by replacing O_i by its equivalent $\sum_{j=1}^{m} O_{ij}$. With these changes the LP (1)-(14) is a bounded variables LP (see [6],[9],[10]) with effectively only 5n+m+4 constraints (3)-(7), (10) and (13)). Since linear programming solution time typically increases with the number of constraints, this is a relatively modest linear program. Thus, many alternative policies can be tried without exceeding a computing budget.

IV THE REGIONAL ASPECTS OF THE MODEL

One of the important features of the model is its focus on the regional economic impact of petroleum allocations. Both the objective function (1) and the supply constraints (10) include regional coefficients of production for each output category. The model does not include regional productions of all inputs, but only for those factors of production which are directly involved in policy considerations. These inputs will be referred to as strategic factors of production and because of their importance in the model will be examined in detail.

IV.1 Regional Employment Coefficients

Regional employment coefficients, e_{ij}, are an integral part of the model. The objective function (1), as well as constraints (13) utilize these coefficients to relate industrial employment and output within each region.

These labor coefficients can be estimated by computing total regional employment within the planning period by a particular industry and dividing it by the output produced during that same period. Output can be measured in final dollars, physical units, or value added. Once this ratio is established, the employment coefficient will be determined.

As an illustration, assume that output in industry i in region k was $\tilde{O}_{ij}$, and the labor required in industry i to produce $\tilde{O}_{ij}$ was $\tilde{E}_{ij}$ (see Figure 1). Further assume an employment function which is linear and passes through the origin. By passing a line between $(\tilde{O}_{ij},\tilde{E}_{ij})$ and zero, we estimate an employment function whose slope is e_{ij}, the employment coefficient.[7]

This procedure assumes that there is not an initial minimum amount of employment required for production of the first unit. At first, this may appear to be unrealistic, but it is doubtful that production would be curtailed to the

---

[7] If whole industries are expected to shut down, then the overhead employment (the employment level required to produce the first unit, e.g. office employees, staff, etc.) must be estimated.

point where an initial fixed employment level would be significant. That is, a more realistic employment function, which had a positive intercept, would only diverge dramatically if output fell far below the normal level.[8]

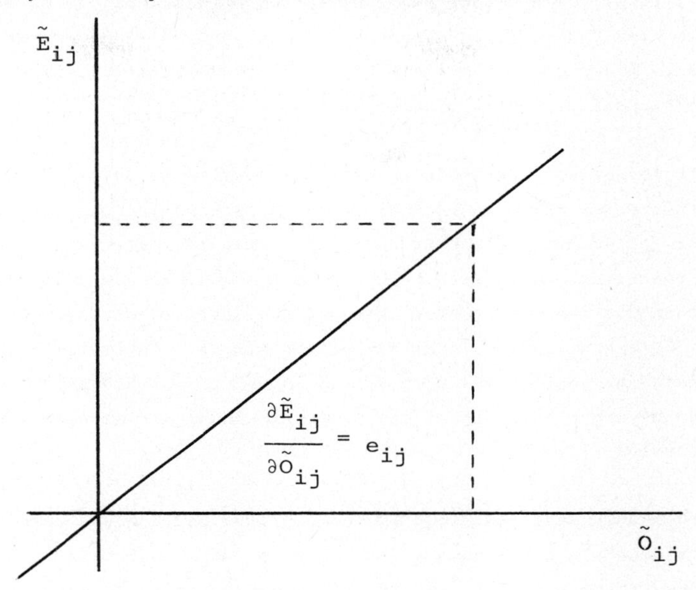

Figure 1

Estimating the Employment Coefficient
in Region k and Industry i

IV.2 Regional Coefficients for Petroleum Inputs

The second set of strategic factors which are considered at the regional level are petroleum products. Regional input coefficients for each petroleum product are included for every category of regional output used in the model. The inclusion of these coefficients will cause a shift in industrial allocations of these petroleum inputs from less efficient to more productive regions within the limits imposed by other constraints in the model. The upper bound on these shifts will be the production capacity for each industry within each region, constraints (11), and the lower bound will be a set of minimum output levels which must be maintained to insure national and regional economic viability, constraints (3), (4), and (12). It is important to realize that the balance of the production mix is maintained, not at the regional level, but at the national level by a complete input-output module in constraints (5).

Regional coefficients for the four petroleum products can be estimated by the same method as described for labor coefficients in the previous section. The

---

[8]The discussion in footnote #1 applies to this point in particular.

industrial output and petroleum consumption data required in the estimation process are not available at the regional level at the present time. This information would have to be collected by sampling the industrial sectors in each state. This method has been successfully employed in interregional trade research, but the coefficients generated did not break petroleum products into fine enough categories to be useful in this model. [4],[5]

It may be more accurate to base the model on a detailed inter-regional input-output model which would include a complete set of input coefficients for all factors of production rather than just these five strategic factors. The work and data required to establish this approach, however, would be considerably greater than the model presented here and the inreased accuracy may make only a marginal difference in the final allocation scheme. The estimation of the regional coefficients for four petroleum products is not an unreasonable task, and a detailed input-output analysis of the national economy is already available from several sources [1].

When the model described in the previous section was developed, the availability of accurate data was a primary consideration. Several other approaches were pursued in detail, but discarded because of their unrealistic data requirements.

One minor problem does arise at the regional level from the indirect use of petroleum products through the consumption of electric power. A major portion of this power is used by industry for production and, therefore, is an indirect use of residual fuel oil. The model treats electrical power as an intermediate good and provides input coefficients in the model's constraint equations (5). The model indirectly determines the amount of electrical power allocated to each industry in each region, and the solution matrix indicates the fuel allocation to the utility industry in each region for generating industrial power. The allocations for residential or other essential uses is allocated prior to the model. The rationing of electrical power to industry may require new technological or administrative methods.

V SUPPLY LEVEL CONSIDERATIONS

Since the allocation LP (1)-(14) is a single period model, the supply level, S_k (k=1,2,3,4), for each refined petroleum product, is the amount available for use in the planning period. This level depends on the quantity which is in inventory at the beginning of the period, the quantity produced during the period and available for consumption in the period, and the inventory required at the end of the period.

If estimates for the supply levels are available, they can be input directly into the model. Various alternative supply levels can be examined to determine their effects on the economy. Since an increase in the production of one refined petroleum product necessitates a decrease in another, the effects of various trade-offs can be analyzed. Model sensitivity analysis will be pursued in more detail in the section on alternative uses and strategies.

Refinery production, to date, has been established, via a refinery LP model, in response to current market prices for the various petroleum commodities. In the past, when supply was not a constraint, no difficulties were encountered. Now, however, in a short supply economy difficulties arise.

If market prices do not reflect societies' real needs, refinery production will be out of phase. As an example, consider the situation where the marginal revenue for gasoline exceeds the original revenue for heating oil, thus encouraging refineries to reduce gasoline instead of heating oil. In winter months such a policy will create a heating oil shortfall. If, as a result, heating oil prices go up, a gasoline shortfall can be expected during the following summer months.

Based on a known limited supply of crude oil, a nonlinear programming model can be employed to determine optimal prices for the different refined products so as to bring consumer demand in line with available supply and to determine optimal production levels to meet this demand. This model can be used to establish the supply levels, S_k, necessary for the allocation LP. It makes use of an aggregate refinery LP model of the petroleum industry and indirect demand relationships for each refined product.[9]

The aggregate refinery LP can be stated as follows:

$$\text{maximize} \sum_{j=1}^{n} (p_j - c_j)x_j \tag{15}$$

$$\text{subject to:} \quad Ax = b \tag{16}$$

$$x \geq 0 \tag{17}$$

where: n is the number of refined petroleum products

x_j is the quantity of refined petroleum product j to be produced

p_j is the price per unit for product j

c_j is the cost per unit for product j

$x = (x_1, x_2, \ldots, x_n)$

---

[9] We have received word that an aggregated refinery LP model has been developed by Bonner and Moore [7]. In addition, research in the determination of demand relationships for the refined petroleum products is underway [8].

The objective is to maximize revenue (15) subject to maintaining the refinery input-output relationships and non-negative production as represented by constraints (16) and (17) respectively.

An indirect demand relationship, $P_j(x_j)$, is depicted:

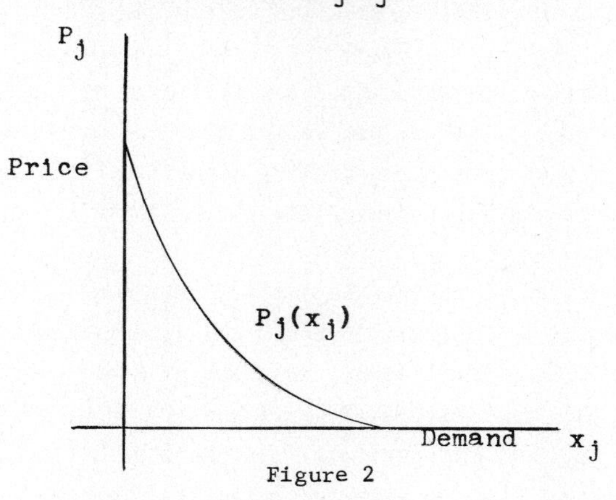

Figure 2

The optimal production, $x = (x_1,x_2,...,x_n)$ and the price levels, $P_1(x_1),P_2(x_2),...,P_n(x_n)$, at which refineries will be motivated to produce at a level where the demand will meet the available supply can be determined by solving the nonlinear program[10] derived from (15)-(17) by replacing (15) with

$$\text{maximize} \quad \sum_{j=1}^{n} (P_j(x_j)-c_j)x_j \tag{18}$$

The following anti-inflation constraint can be added if it is observed that prices might advance too high:

$$P_j(x_j) \leq u_j \tag{19}$$

In addition, the following constraint can be added if the minimum levels of certain products (e.g. fuel oil) must be met:[11]

$$x_j \geq m_j \tag{20}$$

VI ALLOCATION MATRIX

After solving (1)-(14) via linear programming, a solution matrix, A_k, for each of the four refined petroleum classes is determined:

---

[10]The model has a nonlinear objective function with a linear constraint set and as such possesses a special structure which can be taken advantage of.

[11]Notice that constraints (19) and (20) are actually surrogates for one another via the demand relationships. The following constraint would be added when implemented:

$$x_j \geq \max\,(m_j, P^{-1}(u_j))$$

industry i \ region j	1	2	. . . .	j	. . . .	m	Industry Totals
1	A_{11k}	A_{12k}	. . . .	A_{ijk}	. . . .	A_{1mk}	A_{1k}
2	A_{21k}	A_{22k}	. . . .	A_{2jk}	. . . .	A_{2mk}	A_{2k}
⋮	⋮			⋮		⋮	⋮
i	A_{i1k}	A_{i2k}		A_{ijk}		A_{imk}	A_{ik}
⋮	⋮	⋮		⋮		⋮	⋮
n	A_{n1k}	A_{n2k}		A_{njk}		A_{nmk}	A_{nk}
Region Totals	$\bar{A}_{1k}$	$\bar{A}_{2k}$		$\bar{A}_{jk}$		$\bar{A}_{mk}$	

where A_{ijk} is the total allocation of petroleum product k to industry i in region j, $A_{ik} = \sum_{j=1}^{m} A_{ijk}$ is the total allocation of petroleum product k to the ith industry, and $\bar{A}_{jk} = \sum_{i=1}^{n} A_{ijk}$ is the total allocation of petroleum product k to the jth region.

Once allocation A_{ijk} has been determined, allocation to individual firms or plants within industry i in region j can be made via some simple heuristics. For example, allocation can be based on the proportion of a firm's employment (or output) in the region's industry base,

$$A_{pijk} = A_{ijk}(E_{pij}/E_{ij}),$$

where E_{pij}/E_{ij} is the fraction of the total employment of industry i in region j at firm p. The quantity, A_{pijk}, becomes the firm's ration ticket, so to speak.

VII ALTERNATIVE USES AND STRATEGIES

The LP model (1)-(14) is intended to be used as a planning tool. Various policy decisions can be tested to determine their effect on both regional and national employment and output. For example, effects of different minimum output levels for the various vital goods can be established. The disruptions resulting from emphasizing some goods over others can be determined. As mentioned previously, the effects of changing the refinery product mix can be ascertained. Foreign crude purchases can be analyzed on a cost-benefit basis. The outcome of increased capital investment and increased capacity can be estimated. In addition, the impact on the national economy of various regional constraints can be determined.

In addition to finding an optimal distribution, LP models have additional capabilities. Once an optimal allocation has been obtained, the effects of uncertainty can be established via LP sensitivity analysis. Some feeling for how sensitive the model is to estimation error can be ascertained. For example, sensitivity analysis can be used to determine how sensitive an optimal solution is to changes in the employment/output ratio e_{ij} and the consumption/output ratio r_{ijk}. In addition, the linear programming technique is flexible enough to be used to determine how large a change in any parameter is necessary to produce a desired change in the allocation plan.

Parametric linear programming can be employed to determine petroleum requirements if the output of the economy were reduced by a variable percentage. Using this technique, the economy can be pared down incrementally to a point where a feasible supply of petroleum is available for an optimized output mix.

The model can be used in a number of other ways as well. It can be used to allocate a single refined petroleum product, e.g. fuel oil, many refined petroleum products, or it can be used to determine in what proportions crude should be refined into various petroleum products (which are in turn optimally allocated) to maximize employment subject to the numerous economic constraints mentioned earlier.

Finally, the model can be used to evaluate the effects of alternative energy sources (e.g. coal, natural gas, solar heat, etc.). Alternative energy source levels are production inputs in the LP model. Changes in these levels correspond to changes in the input-output coefficients (constraints (5)). An LP sensitivity analysis can determine how sensitive employment and output are to changes in the levels of usage of alternative energy sources.

VIII CONCLUSIONS

This model represents a potentially powerful tool which could be used to generate an optimal oil allocation plan. It has three distinct advantages over any method which has been developed up to this time.

First, it is flexible. It will allow administrators and planners to try various approaches to rationing by manipulation of the constraints on output and regional oil supply. Because it can be computerized, many different alternatives can be tried in a short period of time. The model can be run at different times throughout the winter with as many different planning horizons as the database will allow. Second it has practical data requirements. It primarily uses standard regional economic data which is regularly collected by the Commerce and Labor Departments. The effort required to estimate regional petroleum input coefficients does not seem unreasonable, and any alternative approach would have to employ fuel oil consumption data in some form. Both the machinery and expertise is available to obtain this information. Finally, the model makes explicit assumptions about priorities among users is real terms. Minimum levels of output and oil supply are assigned rather than an artificial indices or rankings. The model deals with actual production and employment requirements rather than a set of surrogate indices and, therefore, is more accurate and realistic.

Because of its realistic approach, practical data requirements, and flexibility, this model deserves serious consideration. It brings a powerful set of analytical and technological tools to bear on one of the most complex and vital problems that the United States has faced in recent times.

REFERENCES

Hannon, Bruce, and Folk, Hugh, 1973. "An Energy, Pollution and Policy Model," Center for Advanced Computation, No. 68, University of Illinois, February. [1]

Isard, Walter, 1960. Methods of Regional Analysis, M.I.T. Press, Chapter 7. [2]

Moore, Craig L., 1973. "A New Look at the Regional Trade Multiplier," Northeast Regional Science Review, Vol. 3, pp. 164-171. [3]

Polenske, Karen, R., The United States MultiRegional Input-Output Model, Heath Lexington Books, (forthcoming). [4]

Rodgers, John M., State Estimates of Interregional Trade Flows, Heath Lexington Books, (forthcoming). [5]

Simonnard, M., 1966. Linear Programming, Prentice-Hall, Inc., pp. 206-211. [6]

Sisson, William, 1973. Personal communication, Marathon Oil Company. [7]

"The Gordian Knot of Gasoline Prices," Business Week , December 15, 1973, p. 23.[8]

Wagner, Harvey M., 1958. "The Dual Simplex Algorithm for Bounded Variables," Naval Research Logistics Quarterly, Vol. 5, pp. 257-261. [9]

Wagner, Harvey M., 1969. Principles of Operations Research, Prentice-Hall, Inc., pp. 148-151. [10]

THE WISCONSIN REGIONAL ENERGY MODEL: A DYNAMIC APPROACH[1] TO REGIONAL ENERGY ANALYSIS

Wesley K. Foell, John W. Mitchell and James L. Pappas

I INTRODUCTION

The economic dislocations occasioned by the rapid changes in energy supply and demand relationships over the past several years have catapulted the energy issue into a prominent position in all planning processes. With the newfound realization of the importance of energy in virtually every aspect of an industrialized society came a recognition of an almost total lack of knowledge concerning the complex interrelationships involved. Because of the intricate manner in which energy is intertwined with virtually all characteristics of an industrialized society, a physical energy system is comprised of an extremely large set of interconnected components. Energy flows are dependent on, among other things, social and economic forces, cultural and demographic characteristics, resource endowment, industrial composition, and levels of saving and income. If energy-related decisions are to be improved in the future, methods for studying the energy system and testing alternative policies must be developed to assist in the decision-making process.

In recognition of the importance of developing a greater understanding of energy systems, a large interdisciplinary group on the Madison campus of the University of Wisconsin formed the Energy Systems and Policy Research Group, devoted to the study of energy systems and policy alternatives from a broad multidisciplinary perspective. The primary efforts of this research group have been aimed at the development of quantitative tools, useful for improving our understanding of energy systems and capable of assisting in more thorough analysis of energy policy alternatives.

A major outcome of this research effort has been the development of a regional energy model for the state of Wisconsin. The model is a computerized dynamic simulation model structured to treat the State's energy system within a multidimensional framework that describes energy demand, conversion, transport, use, and environmental impact, taking account of technological and economic considerations. It is basically an extensive energy information system designed to describe the Wisconsin energy system and its relationship to other characteristics of the State.

---

[1]The research described in this paper was performed by the Energy System and Policy Research Group at the University of Wisconsin under the sponsorship of the National Science Foundation (RANN Division) and the Upper Great Lakes Regional Commission.

The Wisconsin Regional Energy Model has been employed in the analysis of a variety of energy related issues. A few of the specific questions that have been addressed are:

1. What additions will be required in the coming decades to the electricity generating, transmission, and distribution facilities in the State?
2. What are the environmental impacts associated with alternative future energy use patterns in Wisconsin, and what are the feasible trade-offs?
3. What role can energy conservation play in determining the State's energy future and how can such conservation be effected?
4. What would be the impact of primary fuel diversions from the State?

In this paper we provide a general overview of the work being carried on by the Energy Systems and Policy Research Group, including a description of the current structure of the Wisconsin Regional Energy Model. The results of several applications of the model to energy policy analysis are also presented.

II THE WISCONSIN REGIONAL ENERGY MODEL

The WISconsin Regional Energy Model (WISE) is a computerized dynamic simulation model designed to describe the technological-economic-environmental interactions in the Wisconsin energy system. It consists of a collection of submodels, programmed in FORTRAN, which combine in simple mathematical terms data and information about energy flows in Wisconsin to describe or simulate the energy system and its relationship to other characteristics of the State, e.g., demographic, economic, and environmental. A simulation structure was chosen for several reasons. First, simulation is a convenient method of integrating the variety of analytical techniques likely to be employed in a multidisciplinary effort of this type. Second, a simulation structure provides a great deal of flexibility in the modeling process. For example, it enables one to modify selected components of the system without the necessity to rework the entire model and to focus attention on specific areas of the energy system as well as on the system as a whole. Finally, the simulation structure lends itself to the scenario generating approach that is extremely useful in the analysis of major policy issues and alternatives. That is, simulation facilitates the application of the model to questions of the "what if" type.

Although there are numerous ways one could describe the overall structure of the WISE model, one of the more revealing is by component subsystems as illustrated in Figure 1. Using this scheme, the model can be conveniently subdivided into five major components:

1. Socio-Economic Activity Models
2. External Energy Supplies and Prices
3. End-Use Energy Demand Models

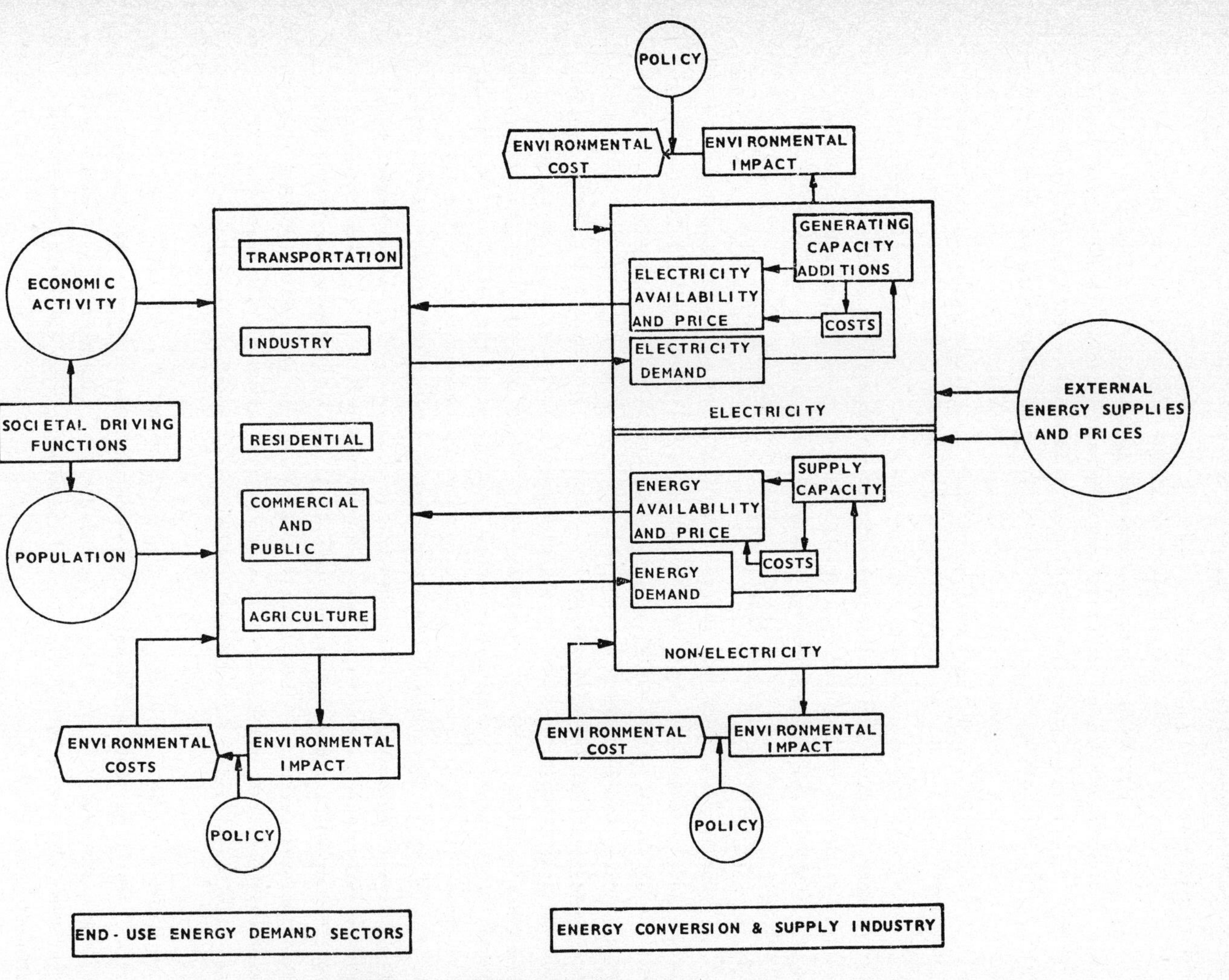

FIGURE 1. WISCONSIN REGIONAL ENERGY MODEL

4. Energy Conversion and Supply Models
5. Environmental Impact Models

Each of these components is described below.

II.1 Socio-Economic Activity Models

This set of submodels provides information on population, demography, and economic activity distributed spatially across the State. Most of the inputs to these models are exogenous; their outputs drive the end-use energy demand submodels.

The population model is an age-, sex-, and county-specific model developed by the demographer's office of the State of Wisconsin. It includes considerations of migration, fertility, and mortality. Using this model, population and demographic projections have been obtained through the year 2000.

The current economic activity component of the WISE Model relies primarily on exogenously determined estimates of economic activity. Specific measures of economic activity employed are value-added in manufacture, wholesale, retail sales, receipts to service industries, and government expenditures. In general, trend techniques have been employed to project these economic variables into the future, with various alternative scenarios created by user-specified modifications of the trends. Current work in this area is aimed at tying the WISE Model to an econo metric model of the Wisconsin economy. In addition to providing improved values for the driving functions, this approach will allow a direct feedback loop to be established between the energy model and the economic activity variables.

II.2 External Energy Supplies and Prices

The State of Wisconsin, having no internal fuel resources, is almost entirely a consumer of energy (the small exception being in the hydro generation of electricity). Because of this, prices and availability of fuels are treated as being exogenously determined in the WISE Model. This component then serves primarily as a data bank which provides information on fuels which could be imported into Wisconsin. This data is stored in the form of time-dependent descriptions of origin, composition, energy content, price, and availability of oil primary fuels.

Routines have been developed which enable the user to either explicitly specify the future prices of various energy sources or to assume any of a variety of relative growth rates for specific fuel usages. This allows one to explore questions of energy substitutions and the impact of resource limits on economic growth.

II.3 End-Use Energy Demand Models

The demand for energy is modeled with a five-sector classification corresponding to major user distinctions in terms of both demographic and economic variables. Each user class provides the basis for a demand submodel which examines the energy consumption relationships within that sector.

Some of the end-use models result from a combined economic-technological approach, whereas with others the economic approach is strictly complementary to a technical or engineering model. The basic structures of each demand submodel are outlined in the following sections.

II.3.1 Transportation Energy Demand Model

In its present stage of development, this model is comprised of three subroutines describing passenger auto transport, intercity mass transit, and freight transport. The basic procedure in each subroutine is to determine the transportation demand in passenger- or ton-miles, using inputs from other submodels. Energy intensiveness factors which depend on the transportation mode and the load factor (average number of tons or passengers per vehicle) determine the energy use from the demand.

The model for energy consumed in passenger auto travel shown in Figure 2 indicates the basic approach used in this demand submodel. Passenger transport demands, in terms of passenger miles, are derived from a demographic model which provides a distribution of trips per household, distributed over family size and automobile ownership--the function TPH in Equation 1, Figure 2--and a distribution of intercity and intracity trip lengths. The other major variable in the passenger automobile energy demand model is an energy intensiveness factor, EI, which is highly dependent upon the vehicle type and year of manufacture. Using variables such as the fraction of a particular type of vehicle entering the population in a given year, its energy intensiveness, and a measure of the likelihood that it will still be in the vehicle population in a given future year, the model computes an effective energy intensiveness coefficient for each year of the simulation. Combining these three factors in the passenger automobile transportation energy demand model provides a mechanism for examining the impact of consumer preferences, technological innovations, and policy decisions affecting the use of private vehicles for passenger transport.

The transportation energy demand model also contains environmental impact subroutines so that pollutant emissions can be computed for any of the hypothesized future vehicle populations and use patterns. The transportation model is described in greater detail in (1).

II.3.2 Industrial Energy Demand Model

The industrial sector of Wisconsin's economy is responsible for approxi-

$$(1) \quad \left[EPH\right]_{i,j,k,t} = \left[TPH(VO_j, FS_k)\right] \cdot \left\{ \left[\int_{X_1}^{R_{i,t}} X\, G_t(X)\, dX \right] \cdot \left[(EI)_u\right]_t \right.$$

AUTO TRANSPORTATION ENERGY IN YEAR t PER HOUSEHOLD OF FAMILY SIZE FS_k AND VEHICLE OWNERSHIP VO_j IN CITY OF POPULATION RANGE i

DISTRIBUTION FUNCTION FOR TRIPS PER HOUSEHOLD AS FUNCTION OF VEHICLE OWNERSHIP AND FAMILY SIZE

MEAN DISTANCE OF TRIP IN CITY OF POPULATION RANGE i AND WITH MAXIMUM TRIP LENGTH R_i

ENERGY INTENSIVENESS OF URBAN AUTO, BTU/VEHICLE MILE

$$\left. + \left[\int_{R_{i,t}}^{X_f} X\, G_t(X)\, dX \right] \cdot \left[(EI)_r\right]_t \right\}$$

MEAN DISTANCE OF TRIP IN RURAL AREA

ENERGY INTENSIVENESS OF RURAL AUTO, BTU/VEHICLE MILE

$$(2) \quad \left[E\right]_t = \sum_i \sum_j \sum_k \left[EPH\right]_{i,j,k,t} \cdot \left[H_i(VO_j, FS_k)\right]_t \cdot N_i$$

TOTAL AUTO TRANSPORTATION ENERGY IN STATE IN YEAR t

NUMBER OF HOUSEHOLDS WITH FAMILY SIZE F_j AND VEHICLE OWNERSHIP VO_j

FIGURE 2. MODEL FOR ENERGY CONSUMED IN PASSENGER AUTO TRANSPORTATION

mately 35 percent of the electricity use and about 20 percent of the state's total energy consumption (2). Industry uses virtually all types of fuels, but primarily coal, oil, natural gas and electricity. These fuels are used in the form of heat for elevating temperatures of materials and generating process steam, or as electricity for lighting, electrical processes, and operating machinery.

Industrial plants use energy for numerous diverse purposes. This energy use can be divided into two general categories. First, fuels are used as a pure product in fixed proportions and the usage level can be calculated for any given process and output. The second category of energy use is not directly related to the output level. This category includes energy uses such as lighting, space heating, and running machines. Energy in this second category is often a substitute for labor, capital, or other material inputs. At present, adequate data are not available to allow evaluation of all the above factors, and consequently the model developed in this study uses aggregated variables to represent the effects of the many variables influencing energy use.

The Wisconsin Industrial Energy Demand Model is based on the premise that energy used by industry is determined by the level of industrial output and the amount of energy used per unit of output. This model implies that, at any given point in time, the energy requirements are linearly related to the output. The various types of manufacture are defined and separated according to the Standard Industrial Classification (SIC) system established by the Bureau of the Budget. The two-digit level of disaggregation is used which results in twenty industry categories. This level of disaggregation allows an explicit examination of a changing mix in the State's industrial sector.

The industrial energy demand model contains two primary variables: (1) industrial output, measured in dollars of value-added (VA), provided by the economic activity model, and (2) the energy intensiveness of production (EI), the amount of energy required per dollar of value added. Energy intensiveness is measured in terms of kilowatt-hours (or kilowatt-hours equivalent) used per dollar of value added. The demand equation relating energy use to value added and energy intensiveness is

$$\text{Energy Demand} = \text{Output} \times \text{Energy per unit of Output}$$
$$E_{i,j,t} = VA_{i,t} \times EI_{i,j,t}$$

where i refers to one of the 20 SIC two-digit industries, j refers to the fuel type, and t refers to the year.

A detailed description of this model, the data used in its construction, and an analysis of alternative industrial energy futures is contained in (3).

II.3.3 Residential Energy Demand Model. The residential energy demand model (4) focuses on one of the society's prime energy consumers, the home. In this model the primary causal factor is the number of Wisconsin households. This variable is provided by the population-demographic model. With the number of households projected exogenously, the modeling efforts in WISE are focussed on determination of annual energy consumption per household.

The energy demand of a household is dependent upon a number of demographic, economic, and technical characteristics. In this model, these characteristics are accounted for through the use of the following four factors:

1. Type of home (urban single family, urban apartment, and rural)
2. Ownership fractions for three "base appliances" (space heating, water heating, and central air conditioning)
3. Ownership fractions for 14 "secondary appliances" (e.g., refrigerator, freezer, dishwasher, stove, T.V., etc.)
4. Annual energy use per appliance

The manner in which the above factors fit into the overall structure of the residential submodel is shown in Figure 3. For a given year, the simulation begins with the determination of a demand for new homes, including that resulting from the creation of new families and from demolition of old homes. The next step in the sequence is the calculation of the distribution among the three housing types. These distributions are determined by past trends or may be specified by the model user according to the scenario under study. Because apartments consume significantly less energy for heating and cooling, the fraction of total new residences that are apartments is an important specification that greatly affects energy use. The fraction of households possessing each of these appliances is forecast for each year. For base appliances, these fractions are tied to housing class and age. In addition to the fuels used for the base appliances in new homes, the model accounts for conversions in older homes; e.g., from gas to electric water heaters. For secondary appliances, the fractions are not treated separately by housing class; the fractions specified are for all homes in the State. A saturation effect occurs in secondary appliances; i.e., there exists a logical limit to the fraction of houses that will eventually obtain a particular appliance. The forecasts for appliance ownership are linked to scenarios desribing future availability and prices of basic fuels and electricity.

The calculation of annual energy consumption involves multiplying the total number of household units having each type of appliance by the annual energy use per appliance parameter. This energy consumption factor is modeled to allow the

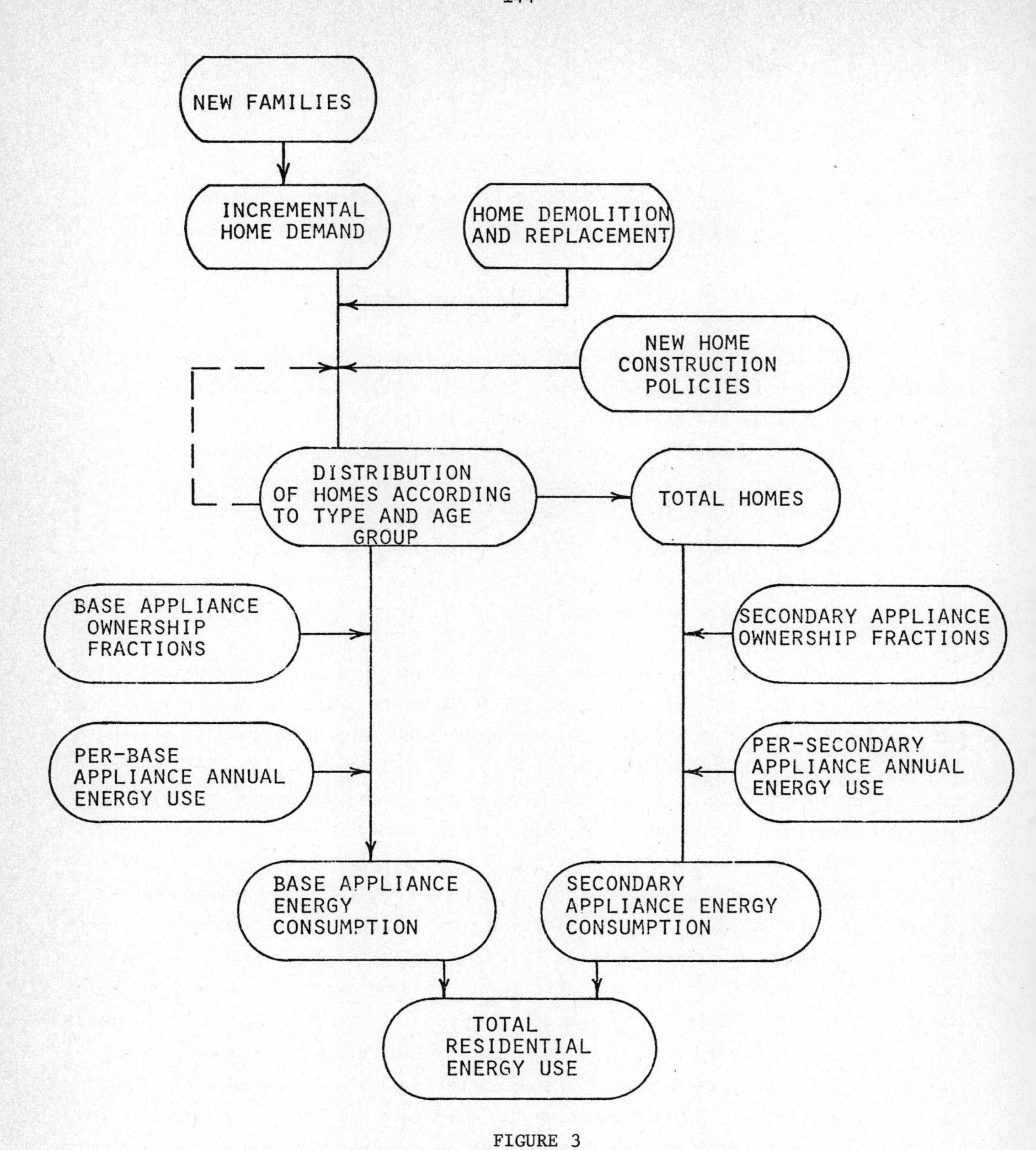

FIGURE 3

RESIDENTIAL ENERGY DEMAND MODEL STRUCTURE

user to specify a wide variety of future appliance efficiencies and use patterns.

II.3.4 Commercial and Public Sector Energy Demand Model. Our research on commercial and public sector energy demand is characterized by two complementary approaches, one of them tightly linked to economic activity; the second based upon an engineering design procedure.

In the first model, energy demand is derived from correlation linking it to commercial sales and government expenditures. The correlations between spending and energy are defined by so-called "energy intensiveness" coefficients which are similar to those used in the industrial submodel.

$$\left.\begin{array}{l}\text{Energy Intensiveness}\\ \text{for Commercial Sales}\end{array}\right\} = \frac{\text{BTU's (or KWH) of energy consumed}}{\text{Dollars fo commercial sales}}$$

$$\left.\begin{array}{c}\text{Energy Intensiveness}\\ \text{for Government}\\ \text{Expenditures}\end{array}\right\} = \frac{\text{BTU's (or KWH) of energy consumed}}{\text{Dollars of government expenditures}}$$

Future annual energy demands are forecast from the product of population, per-capita sales and expenditures, and the appropriate energy intensiveness coefficients. In equation form, $EC_{j,t}$ the total consumption of fuel j in the commercial and public sector in the year t is

$$EC_{j,t} = POP_t(CSPC_t \times EICS_{j,t} + GEPC_t \times EIGE_{j,t})$$

where $CSPC_t$ and $GEPC_t$ are the per-capita commercial sales and government expenditures, and $EICS_{j,t}$ and $EIGE_{j,t}$ are the corresponding energy intensiveness values for fuel j in the year t. Trends in the intensiveness values are derived from historical data or specified by the user according to assumptions about factors such as fuel availability, price, substitutability, etc.

While the above model has provided some broad perspectives on future energy use, it proved to be limited in its capability to answer a number of important questions. Consequently, energy demand in the commercial sector is also being modeled according to the physical processes through which most of the end-use occurs, namely, space heating, space conditioning, illumination and commercial processes. This approach has resulted in the development of an alternative commercial model, relying heavily upon an engineering design approach to arrive at the parameters and functional relationships which characterize energy intensities per square foot of floor area in commercial and public structures. Resultant energy consumption by end-use and fuel type is determined by the product of calculated energy intensities and total commercial floor area assignable to each element of the end-use/fuel type matrix.

Both of the commercial and public sector energy demand models are described in (5).

II.3.5 Agricultural Energy Demand

For energy demand modeling purposes, the agricultural sector is divided into two components - crops and livestock, reflecting the different trophic levels occurring within agriculture. The energy demand is further divided into two categories - direct and indirect energy use. Direct uses are those for which the end-use occurs within Wisconsin, e.g., gasoline to operate a Wisconsin tractor. An example of indirect energy is the energy embedded in chemical fertilizer produced outside of Wisconsin but applied to land within the state. Several forms of indirect energy are tabulated and displayed by the model, but not included in Wisconsin's total energy usage as forecast by the WISE model.

The two primary variables in the crop sector are acres of each crop type and inputs to each acre of crops. Crops and inputs included in the submodel are listed in Tables 1 and 2. Based upon historical data and trends, or modified according to specifications of the user, crop acreage and intpus are forecast through the simulation time period to provide annual energy demands by fuel for each crop.

Table 1

Crops Included in the Crop Submodel

1. Corn for Grain	7. Soybeans
2. Corn for Silage	8. Wheat
3. Alfalfa	9. Rye
4. Other Hay	10. Barley
5. Oats	11. Pasture
6. Tobacco	12. Vegetables

Table 2

Inputs to Land in Crop Submodel

1. Nitrogen Fertilizer	7. Crop Drying
2. Phosphorus Fertilizer	8. Seeds
3. Potassium Fertilizer	9. Labor
4. Insecticides	10. Gasoline
5. Herbicides	11. Electricity
6. Irrigation	12. Diesel

The livestock submodel uses a linear programming approach to calculate annual livestock (dairy and beef) populations in Wisconsin, subject to one or more constraints related to the following factors:

1. Total digestible nutrients (TDN) available
2. Fuel availability
3. Milk marketed

4. Live shipment of dairy animals from the state
5. Beef cow population

Available TDN is an input from the Crop Submodel, and provides a link between the two models. The other four constraints can be trended over time as specified by the user. The livestock population is determined for each simulation year and combined with the appropriate input factors, e.g., gallons of gasoline consumed per dairy cow, to give the total inputs. As in the crop submodel, these inputs are then converted to units of energy to provide a total energy demand from the livestock subsector.

II.4 ENERGY CONVERSION AND SUPPLY MODEL

As indicated in Figure 1, energy prices and availabilities for the primary fuels originating outside Wisconsin are treated as time dependent exogenous functions. Since Wisconsin imports virtually all of its fossil and nuclear fuels, the dynamics of the supply of these fuels is not being modeled. Consequently, the energy conversion and supply component of the WISE model serves primarily a bookkeeping function for these primary fuels. It distributes them to the end-users, the electric utilities, or other conversion industries as dictated by a particular energy use scenario. This bookkeeping does, however, include the storage and transportation systems; e.g., pipeline capacity, and the time lags inherent in their expansion. Consequently, the only Wisconsin energy supply industry currently being modeled in detail is the electricity supply industry.

II.4.1 Electricity Supply Model

The electricity supply industry is being modeled in two separate components: 1) an electricity transmission and distribution submodel, and 2) an electrical generating capacity submodel.

The Electricity Transmission and Distribution Model provides an overall picture of future costs for these two processes and a spatial description of their relationship to land use in Wisconsin. The model, still in an early stage of development, uses either historical trended relationships between generation capacity and transmission line requirements, or hypothetical geometrical land-source patterns to forecast the transmission system requirements for various alternative energy modeled on the basis of patterns of electricity-use density by county.

The electrical generating capacity submodel provides a basis for examining the effect of various plant expansion strategies. As shown in Figure 1 and described below, it forms a focal point for interaction between end-use electricity demand on the one hand, and electricity-related environmental impacts on the other.

The model has three major functions, shown schematically in Figure 4.

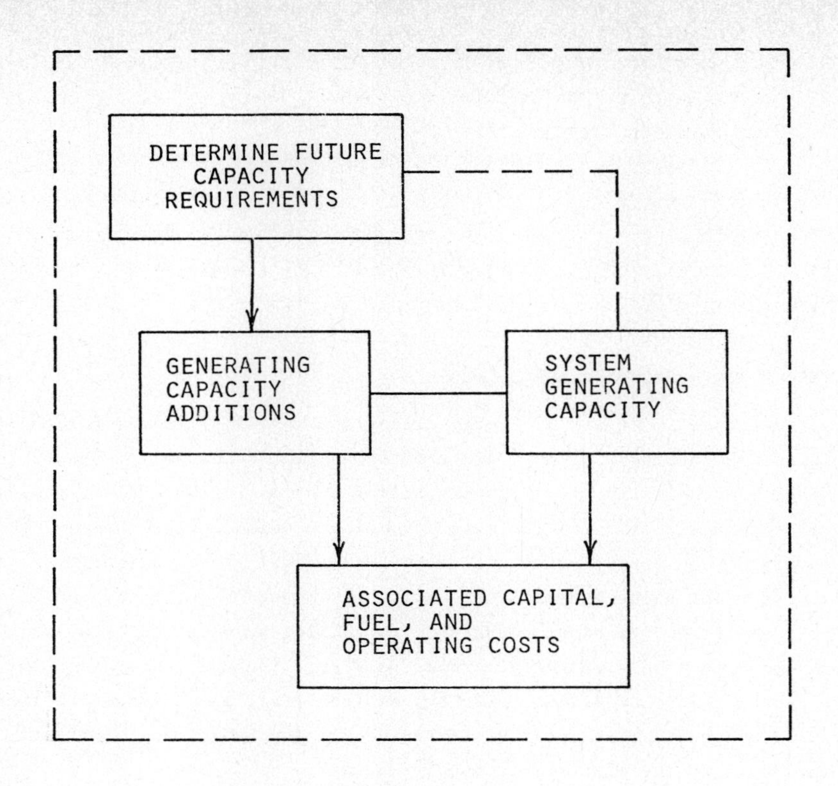

FIGURE 4

ELECTRICAL GENERATING CAPACITY SUBMODEL

1. to convert kilowatt-hour demand requirements coming from the end-use demand sectors into equivalent generating capacity requirement using load duration analysis;
2. to simulate the satisfying of these requirements by adding capacity to the system, dependent upon fuel availability, prices, and technology;
3. to calculate total costs, including capital costs associated with both new and existing generating capacity, the appropriate fuel costs and operating costs over the simulation time period.

The current version of this model is described in (6) and (7). An overview of its structure can be gained from Figures 5-8, taken from those references. Figure 5 shows schematically the general procedure for determining future generating capacity requirements. Capacity utilization factor information is derived from a number of sources, including historical data and postulated trends. The load duration curve is specified by the model user. It can be based on historical data or modified to reflect the changing nature of electricity demand. The user also has the option to alter the shape of the curve during simulation.

Plants are retired according to the general scheme illustrated in Figure 6. The purpose of the scheme is to simulate the process of down-grading capacity to reserve (standby) status or to peaking status. Year of initial operation and expected plant lifetime determine when a particular unit will be transferred between categories or taken off line.

Capacity is added in the general manner shown in Figure 7. Three options are available to aid the model user in investigation of expansion strategies: 1) an external option in which the user explicitly specifies the detailed strategy sequentially during the simulation period; 2) an internally-programmed or pre-set expansion strategy, e.g., a specified time-dependence of the nuclear fraction in the plant mix; and 3) a combination of 1 and 2.

Cost calculations for the different types of generating facilities are performed over the plant lifetimes, using projected long-term time-series of direct and indirect costs, fuel costs, and operation and maintenance costs, including the internalized environmental protection course. The procedure is quite flexible for calculating the costs resulting from alternative capacity expansion strategies. The major variables specified by the user are:

1. the plant capital cost trends, in dollars per kilowatt, which represent projections of expected future capital costs;
2. the fuel costs trends (in cents per million BTU's) which are required for coal, oil and gas fossil fuels, and for Light Water Reactor, High

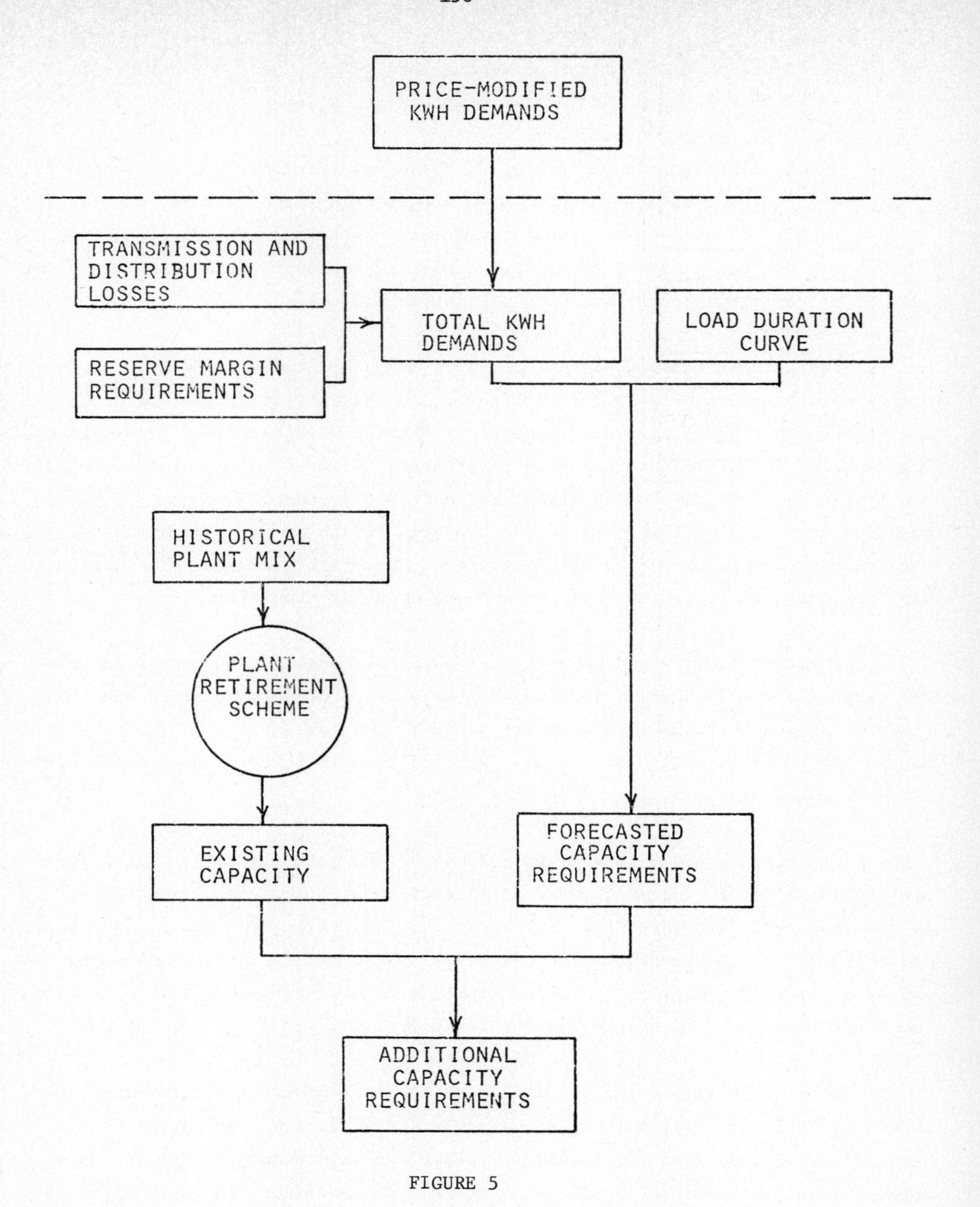

FIGURE 5

CAPACITY REQUIREMENT ROUTINE

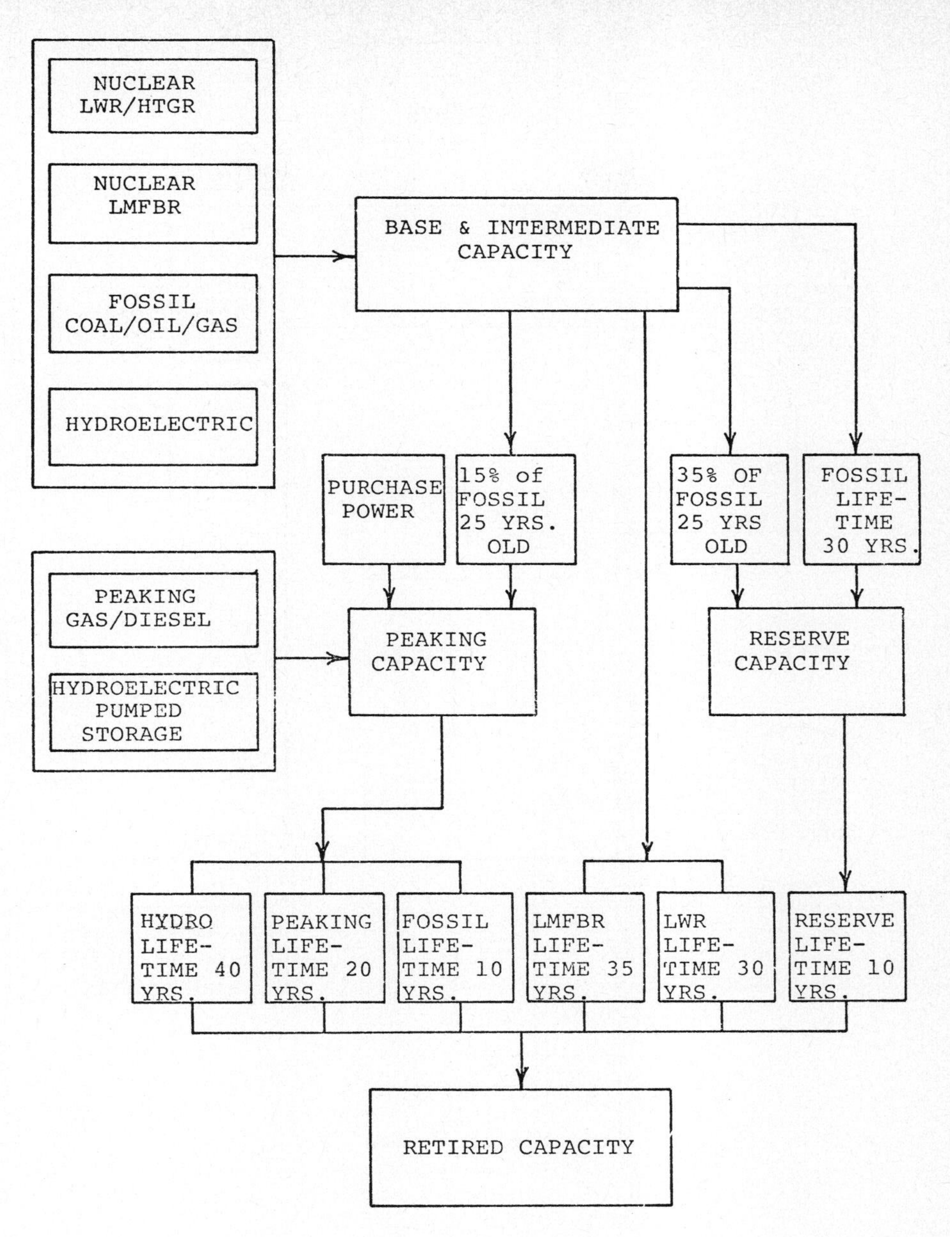

FIGURE 6

PLANT RETIREMENT SCHEME

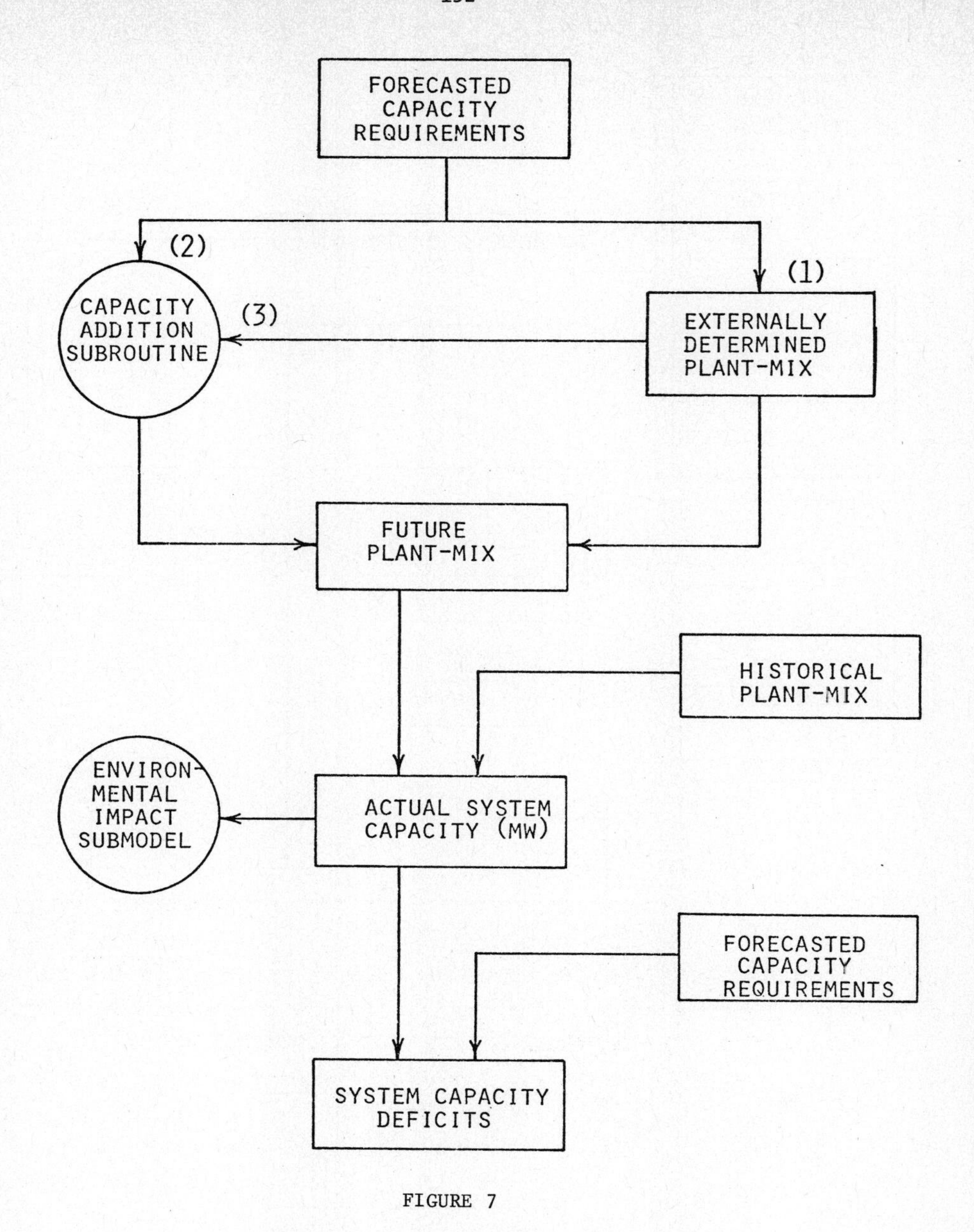

FIGURE 7

CAPACITY ADDITION PROCEDURES

Temperature Gas-Cooled Reactor, and Liquid-Metal Fast Breeder Reactor nuclear fuel loadings;

3. operation and maintenance costs, and
4. adjustment costs, which include cost adjustments for methods of cooling, near-zero rad-waste disposal and sulfur dioxide removal.

Currently under development is an electricity pricing model which will introduce prices more explicitly in the end-use energy demand models. In that model total generation costs will be combined with those provided by the distribution and transmission submodels to give total electricity production costs. As shown schematically in Figure 8, these are converted into electricity prices within the framework of a pricing schedule and a demand scenario specified by the model user. These prices are then compared with the prices used in the end-use demand submodels. If the prices differ, the end-use demand is adjusted so that it is consistent with the forecasted prices. The entire procedure is repeated until the forecasted prices and the end-use demands are consistent, i.e., until an equilibrium is reached between production costs and demand prices.

II.5 ENVIRONMENTAL IMPACT MODELS

The objective of the Environmental Impact Submodel is to simulate on a year-to-year basis the total environmental impacts--to the degree quantifiable--arising from energy usage in Wisconsin (8). For purposes of this model, environmental impacts are broadly defined to include effects on land, air, water, structures, and living things. Included within the definition are not only the health and safety of the general public but also the health and safety of those people employed throughout the energy system. Environmental impact is treated explicitly in <u>all</u> phases of the energy flow system, starting with the fuel extraction phase, e.g., at the oil well, up through end-uses, e.g., emissions from autos. Thus, an important distinction between this component and others in the model is that it treats effects (in this case environmental) both <u>within</u> and <u>outside</u> Wisconsin's boundary.

For the purpose of treating environmental impact, the energy conversion and supply industry can conveniently be divided into electrical and non-electrical categories. Environmental impacts associated with electrical generations are calculated by the electricity impact submodel (9). While it is difficult to display in a general fashion the ways electricity use results in final impacts, Figure 9 shows the pathways for the majority of effects. Final impact as used here is the quantitative result that has a minimum of value-judgment associated with it. Pathway I includes impacts such as air pollution from coal-fired plants, radioactive releases from the nuclear reactor, chemical releases from the power plant,and waste heat. The direct effects of electrical generation shown as Pathway 2 are effects at the power plant such as land use and water use. Pathway 3 accounts for occupational health and accident risk, such as uranium mining accidents and uranium miner's ex-

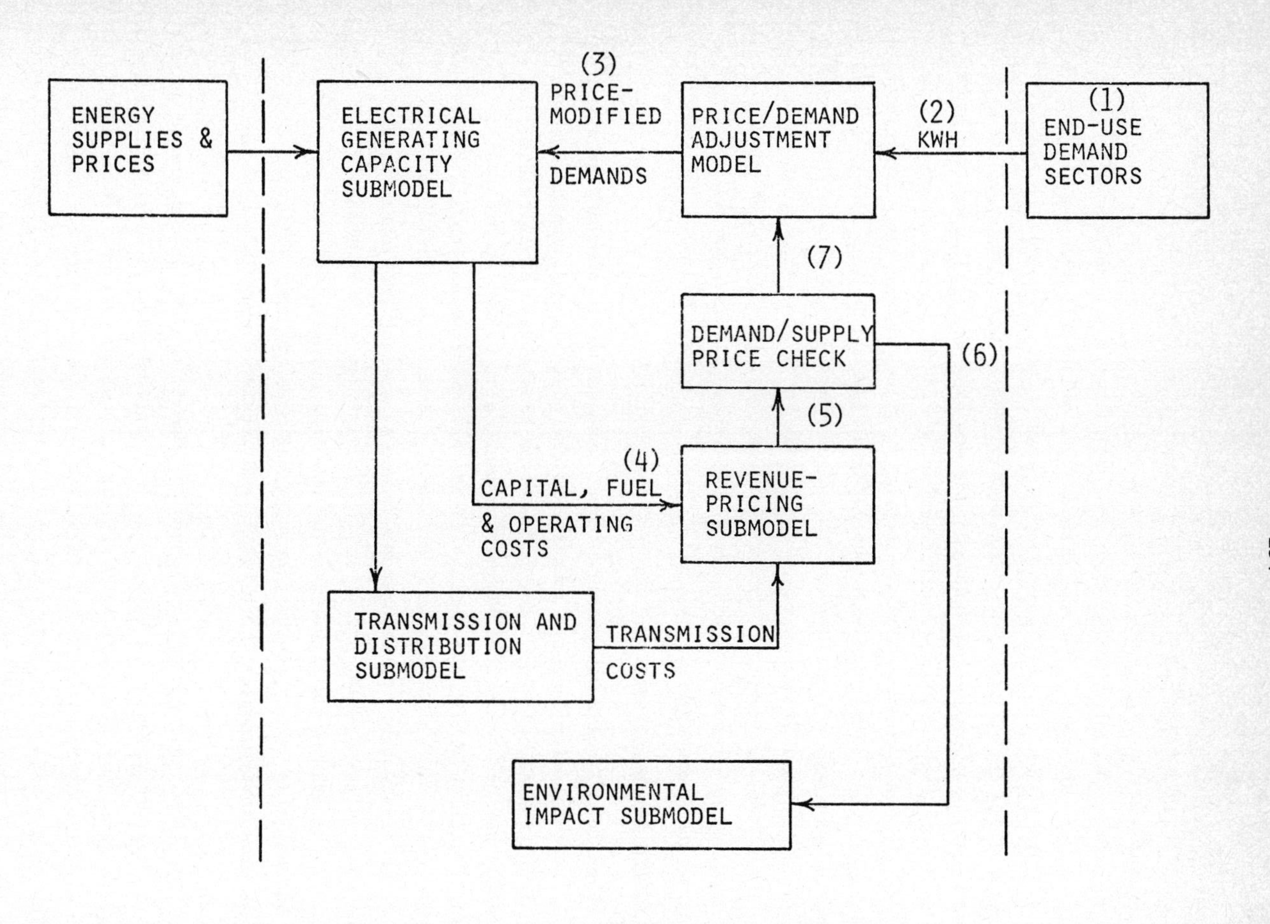

FIGURE 8

ELECTRICITY SUPPLY-DEMAND INTERFACE

posure to radiation. Pollution from fuel cycle operations, such as radioactive releases from nuclear fuel reprocessing plants, are represented by Pathway 4. Occupational health and accident risk at the power plant itself is shown as Pathway 5.

The method for determining the impacts in the electricity impact submodel is quite simple in theory. "Impact factors" are associated with each type of electrical generating capacity. The factors have been determined from collection and analysis of relevant data; they are stored in the model as a function of kilowatt-hours (kwh) generated or capacity (kilowatts) for each type of generating system. Most of the factors are related to electrical generation, e.g., nuclear reactor personnel radiation exposure (manrem per kwh), but some are directly associated with capacity, e.g., land used for cooling systems (acres per kilowatt). The current version of the submodel has five alternative non-peaking electrical generating systems. They are: (1) coal, (2) pressurized water reactor (PWR), (3) boiling water reactor (BWR), (4) high temperature gas-cooled reactor (HTGR), and (5) liquid metal fast breeder reactor (LMFBR). Other advanced technology systems are being added to the model.

The impacts are calculated by multiplying the matrix of impact factors times the generation or capacity of each electrical energy source in each year as shown in the equation:

$$QEI_{i,j,k} = IF_{i,j,k} \times E_{k,j}$$

$$\left\{\begin{array}{l}\text{Quantified} \\ \text{Environmental} \\ \text{Impacts of} \\ \text{type i in year} \\ \text{j resulting from} \\ \text{electrical generation} \\ \text{source k}\end{array}\right\} \quad \left\{\begin{array}{l}\text{Impact} \\ \text{Factor of} \\ \text{type i in} \\ \text{year j for} \\ \text{energy} \\ \text{source k}\end{array}\right\} \quad \left\{\begin{array}{l}\text{Electrical gen-} \\ \text{eration or capa-} \\ \text{city for energy} \\ \text{source k in year} \\ \text{j}\end{array}\right\}$$

These can be summed over the desired number of years to obtain cumulative impacts. They can also be aggregated for all energy sources to get totals for a single year, or for impacts with similar units, e.g., fatalities caused by accidents and fatalities caused by health risks.

The structure of the submodels for treating non-electrical impacts is similar to that shown in Figure 9. The three primary fuel flows modeled are coal, petroleum and natural gas. The impacts modeled are those associated with the end-use of the energy and those associated with the energy supply and conversion system. It includes, for example, impacts associated with emissions from autos, as well as those resulting from extraction, processing, and transport of the petroleum fueling the autos.

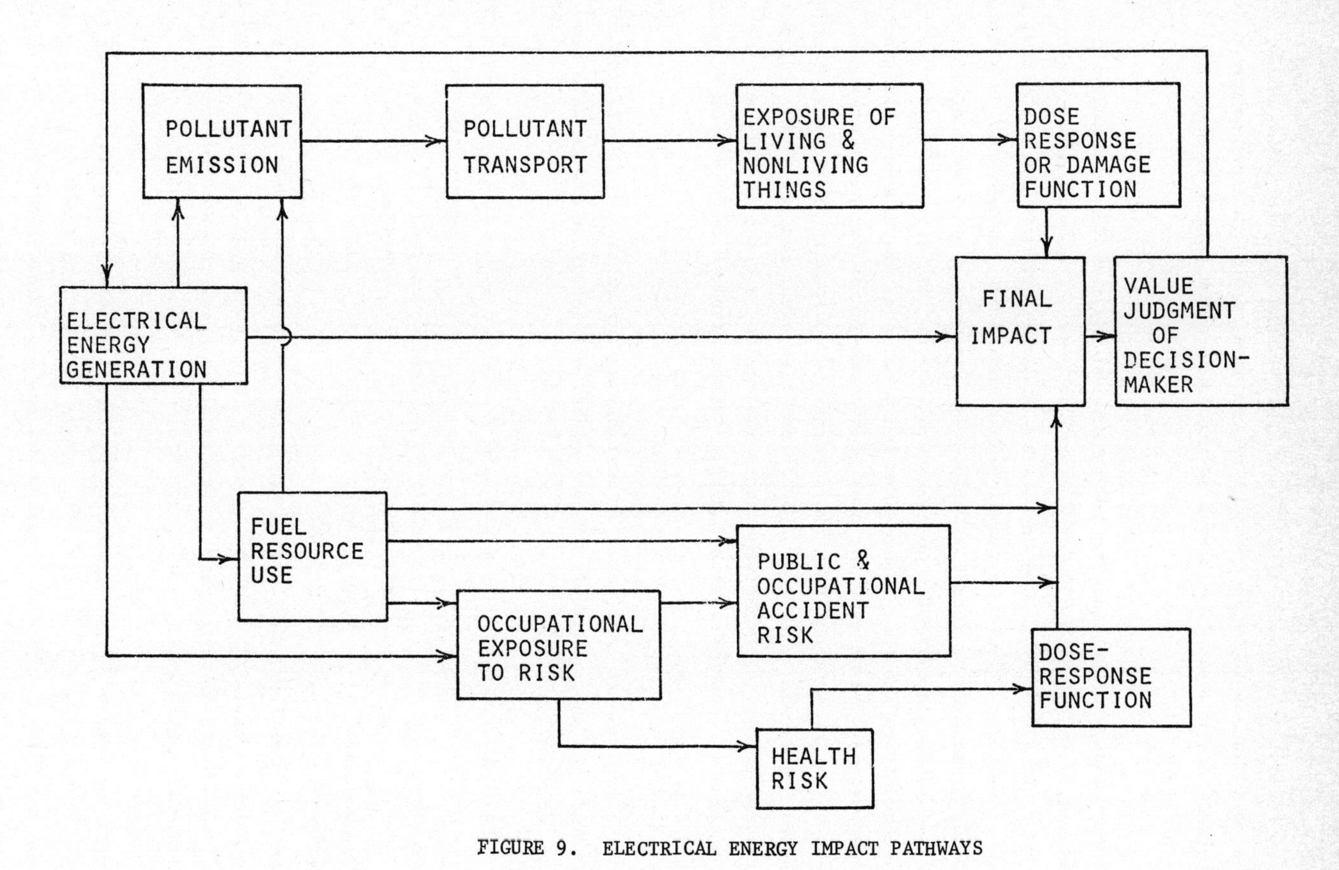

FIGURE 9. ELECTRICAL ENERGY IMPACT PATHWAYS

Although this set of submodels is still in an early stage of development, its general structure has been laid out and detailed modeling is in progress. The environmental impacts associated with end-use of energy will be tied to the end-use energy demand models and/or the physical process models, e.g., the transportation model. In contrast to the impacts of electrical generation, these impacts are quite dispersed throughout the state, necessitating a more detailed model of their geographic distribution and relationship to population locations. An example of this spatial question is the case of health impact from energy-related air pollution. Two complementary approaches to this spatial problem are being followed. The first is to develop correlations between energy-use indicators and ambient pollutant concentrations, e.g., between value-added in manufacture for a given county (or SMSA) and particulate concentrations in that area. A second approach which we are assessing uses county-wide sector by sector economic activity as its driving force and transforms these into pollutant emissions through the use of energy-intensiveness coefficients and pollutant emission factors for the major processes in those economic sectors.

III OPERATION OF THE ENTIRE WISE MODEL

The preceding section has given a brief overview of the structure of the WISE Model and its component subsystems. Before examining several applications of the model to energy analysis, it will prove useful to describe the work that has been devoted to the development of a command language for the simulation routines.

The high degree of interdependence among the various components of the Wisconsin energy system makes it desirable to formally link these components in the model and simulate all parts of the system simultaneously. Further, a prerequisite for a usable simulation model is that it provide for a convenient interaction between the user and the computer. With this in mind, we have developed a computer simulation command language called WISSIM (WISconsin SIMulation) (10, 11). This language allows the user of a model to simulate over a specified time and to modify variables, parameters, functions, etc., to reflect changing conditions or policies in the energy system being studied. Its design requires only that the user know enough FORTRAN to program his model for one year of simulation; WISSIM then provides the user with the needed time-incrementation, graphing, outputs, storage of results, etc. The use of this command language makes it easier for researchers with little or no computer training to integrate their submodels and data into the WISE Model and for policy analysts to interact with the model in evaluating energy alternatives.

Another essential part of the efforts to make the models usable is our interactive computer graphics terminal. Information from the models can be displayed on this terminal in a dynamic optical form, a display which has a powerful impact since it can be rapidly assimilated by both quantitatively- and qualitatively-trained individuals. We believe it is an extremely powerful tool for demonstrating alternative energy futures to a broad spectrum of viewers, including students, the public, and energy decision-makers.

IV APPLICATIONS OF THE WISE MODEL

An initial version of the model was put into operation late in 1973 and is now being applied to the description of alternative energy futures for Wisconsin through the year 2000. Within the context of these alternative findings, a number of specific energy policy questions have been addressed. Several of these are discussed in this section.

IV.1 Long Range Electricity Demand Forecasts for Wisconsin

The forecasting of energy demand is one of the most important aspects of long-range energy planning. Because of the increasingly long lead times required for the installation of energy facilities, a growing need exists for energy forecasts covering periods of one, two, and perhaps even three decades into the future. Demand forecasting plays a particularly important planning role in the case of electricity generation plants because of their long construction lead times and requirements for massive capital investments. Because of the many possibilities of substitution of electricity for primary fuels, a narrowly-based approach to electricity demand forecasting is no longer adequate for predicting future generation requirements.

In response to the need for electricity demand forecasts as a prerequisite for the installation of additional generating capacity in Wisconsin, the WISE model has been used to provide Wisconsin energy forecasts under a wide range of assumptions about causal factors such as population, the nature of economic growth, consumer preferences, primary fuel availability and energy conservation and management measures. This section describes the application of WISE to these forecasts and presents the results in terms of alternative electricity futures for the state. A detailed statement of the assumptions used in these forecasts and subsector energy demand estimates are contained in (12). Although a broad spectrum of futures was studied, only six are presented here. The six electricity futures are:

Case 1 Present Trends
Case 2 High
Case 3 Low
Case 4 Present Trends with Implementation of Energy Conservation
Case 5 High with Implementation of Energy Conservation
Case 6 Low with Implementation of Energy Conservation

The label, Present Trends, is used to signify that the assumptions for those cases involve continuation of the present trends of the causal factors that determine electrical demand. Present Trends does not signify trending of the total electricity demand. The High and Low Cases are based on sets of assumptions that tend to cause the electrical demand to be high and low, respectively. The three cases concerned with conservation are included to indicate the magnitude of potential savings that are associated with selected conservation measures. No attempt has been made to determine the maximum amount of conservation possible in any of the cases. Cases 4, 5, and 6 result from adjustments of Cases 1, 2, and 3, respectively, to account for the selected conservation efforts.

It is important to note that these forecasts are for annual electricity generated and not required capacity. Forecasting of the latter under changing patterns of consumption requires a detailed understanding of the time-dependence of the future components of electricity demands. Although the electricity generating capacity submodel does make use of load duration curves, the curves are not formally derived from the demand models in the current version of the model. Further research is in progress in this area.

IV.1.1 Residential Sector

The primary variations which led to the differing residential energy forecasts presented here included the following:

1. New Homes: For the Present Trends and High Cases, 29 percent of all new residences were assumed to be apartments, while an 86 percent figure was used in the Low Cases.
2. Base Appliances: The percentage of new homes with electric base appliances (heating, cooling, and water heating systems) were as shown in Table 3. Essentially, all new homes were assumed to be all electric with central air conditioning for the High Case.
3. Secondary Appliances: Secondary appliance ownership was assumed to move from the 1970 level to a modest saturation level in the year 2000 for the Present Trends and Low Cases. A 100 percent saturation level by 1990 was assumed for all appliances except washers (87%), electric dryers (59% by 2000), and electric ranges (76% by 2000) in the High Case.
4. Conservation: Conservation measures assumed in Cases 4-6 were: (1) that the economic optimum insulation is installed in electrically heated houses built after 1975, and (2) that room air conditioners purchased after 1975 use electricity 40 percent more efficiently than the average conditioners in use today.

Table 3

Assumed Appliance Ownership Percentages for New Homes

	Urban Single Families	Apartment	Rural Single Families
Present Trends (Cases 1 and 4)			
Electric Heat	12.4	31.0	5.0
Water Heat	62.0	80.0	90.0
Central Air Conditioning	20.0	80.0	5.0
High (Cases 2 and 3)			
Electric Heat	99.9	99.9	99.9
Water Heat	99.9	99.9	99.9
Central Air Conditioning	99.9	99.9	99.9
Low (Cases 3 and 6)			
Electric Heat	12.4	16.0	5.0
Water Heat	38.0	20.0	70.0
Central Air Conditioning	15.0	60.0	10.0

The resulting Wisconsin residential electrical energy consumption forecasts are shown in Figure 10. The range in the year 2000 is more than a factor of two, from a low of 18.8 billion kilowatt-hours in Case 6 to a high of 39.1 billion kilowatt-hours in Case 2. The Present Trends and Low Cases are relatively similar. The High Case stands out because of the extreme assumptions concerning the use of electric heating and secondary appliances.

IV.1.2 Industrial Sector

The six different industrial electricity futures are intended to show the range of possible energy demands with varying assumptions about technological, economic and social conditions. Case 1, the Present Trends Case, is based on a continuation of the past 10 to 15 year trends in the causal variables of the model. Industrial value-added per capita is assumed to grow at the same rate as during the period 1958-1970. This period covers a wide variation in economic trends. Electricity intensiveness growth rates are based on 1954-1967 trends. No leveling off of intensiveness growth is assumed and no limits are set on intensiveness levels. This case represents an intermediate projection. Each of the 20 SIC industries has a specific value for these variables.

For Case 2, the High Growth Case, value-added per capita growth is set at the 1958-1967 rate and as a further floor beneath allowable growth rates, no industry is allowed an annual growth rate of less than one percent. The years 1958-1967 were high growth years for most industries. Therefore, this case implies that very good economic conditions will prevail through the year 2000. Electricity intensiveness growth rates are the same as those used in the Present Trends

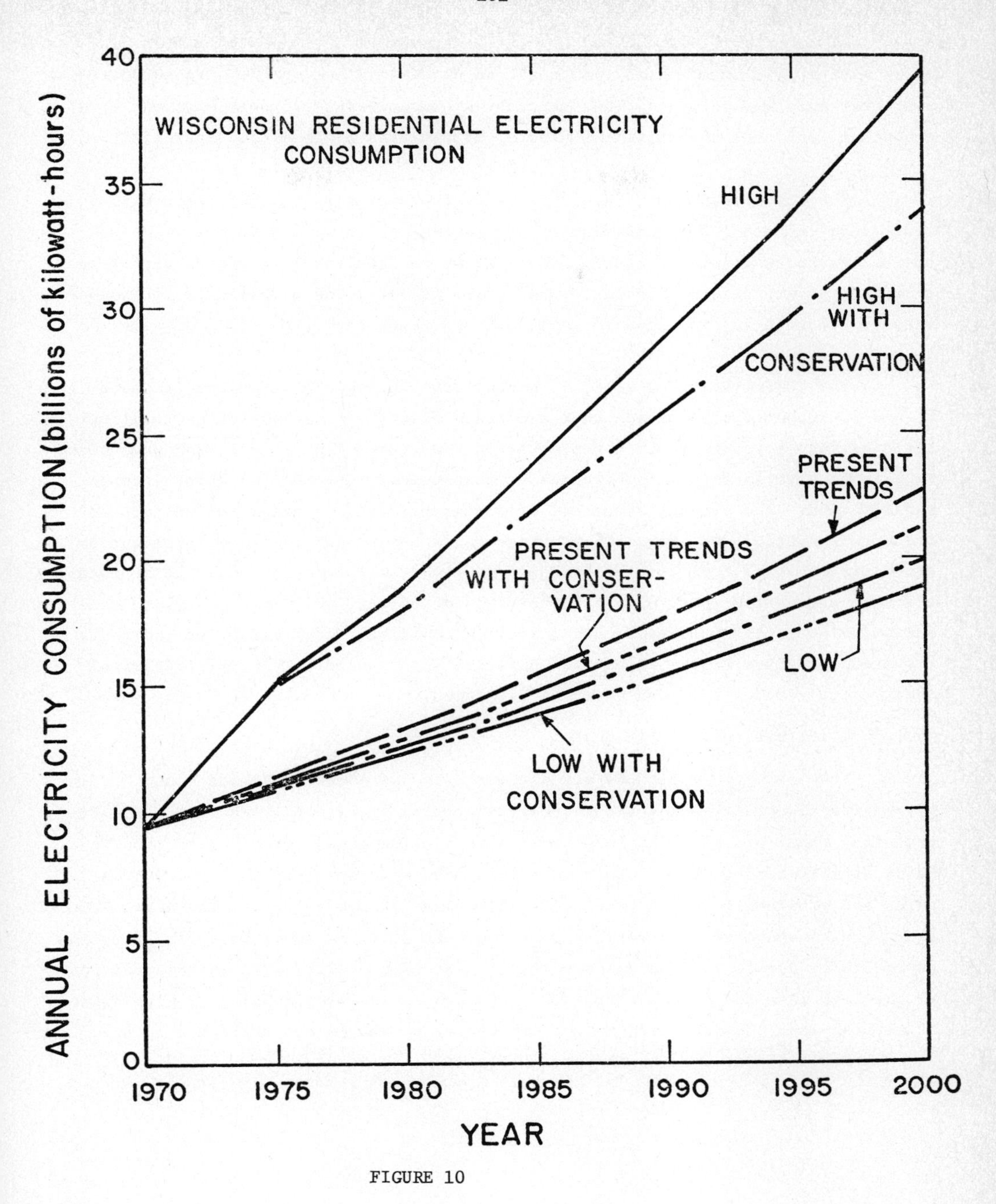
WISCONSIN RESIDENTIAL ELECTRICITY
CONSUMPTION
HIGH
HIGH
WITH
CONSERVATION
PRESENT
TRENDS
PRESENT TRENDS
WITH CONSER-
VATION
LOW
LOW WITH
CONSERVATION
ANNUAL ELECTRICITY CONSUMPTION (billions of kilowatt-hours)
40
35
30
25
20
15
10
5
0
1970
1975
1980
1985
1990
1995
2000
YEAR

FIGURE 10

Case, except a minimum annual growth of one percent is set if the past growth rates are lower than one percent. This case, by assuming very favorable economic conditions and continued intensiveness growth, illustrates a plausible upper limit.

The Low Growth Case, Case 3, assumes that value-added per capita will be lower than in Case 1 and that electricity intensiveness will level off. Value-added per-capita growth is assumed to be equal to the 1958-1970 trends except that a maximum growth rate of 2.5 percent annually is set. The electricity intensiveness growth rate is projected to level off and reach zero growth in 10 years.

The Conservation Cases 4, 5, and 6, have the same assumptions as Cases 1, 2, and 3, respectively, except that a certain percentage is subtracted to illustrate savings possible through conservation measures. Many studies and some plant experience have shown that substantial amounts of energy could be saved through conservation. The present structure of the model does not treat the details of these conservation measures. The savings shown here resulting from conservation are equal to 5% of total use by 1980 and increase to 10% in 1990 and 15% of total electric use by the year 2000. This savings potential, which is probably a conservative estimate, was based on information contained in an Office of Emergency Preparedness study (13). Figure 11 shows the alternative industrial electrical energy requirements produced by these six cases.

IV.1.3 Commercial and Public Sector

The Assumptions for the six alternatives are shown in Tables 4 and 5. The per-capita commercial sales and per-capita government expenditures in the Low Cases gradually taper off until in 1990 growth stops. The annual growth rate then remains at zero through 2000. Note that this does not mean that total commercial sales and government expenditures will not change from one year to the next. The growth rates shown are per-capita, so there will still be an increase in sales and expenditures due to increasing population. The high growth rates in intensiveness in the high case could result from factors such as low availability or high prices of natural gas. Conservation in the commercial sector is accounted for by the same methods as used for the industrial sector.

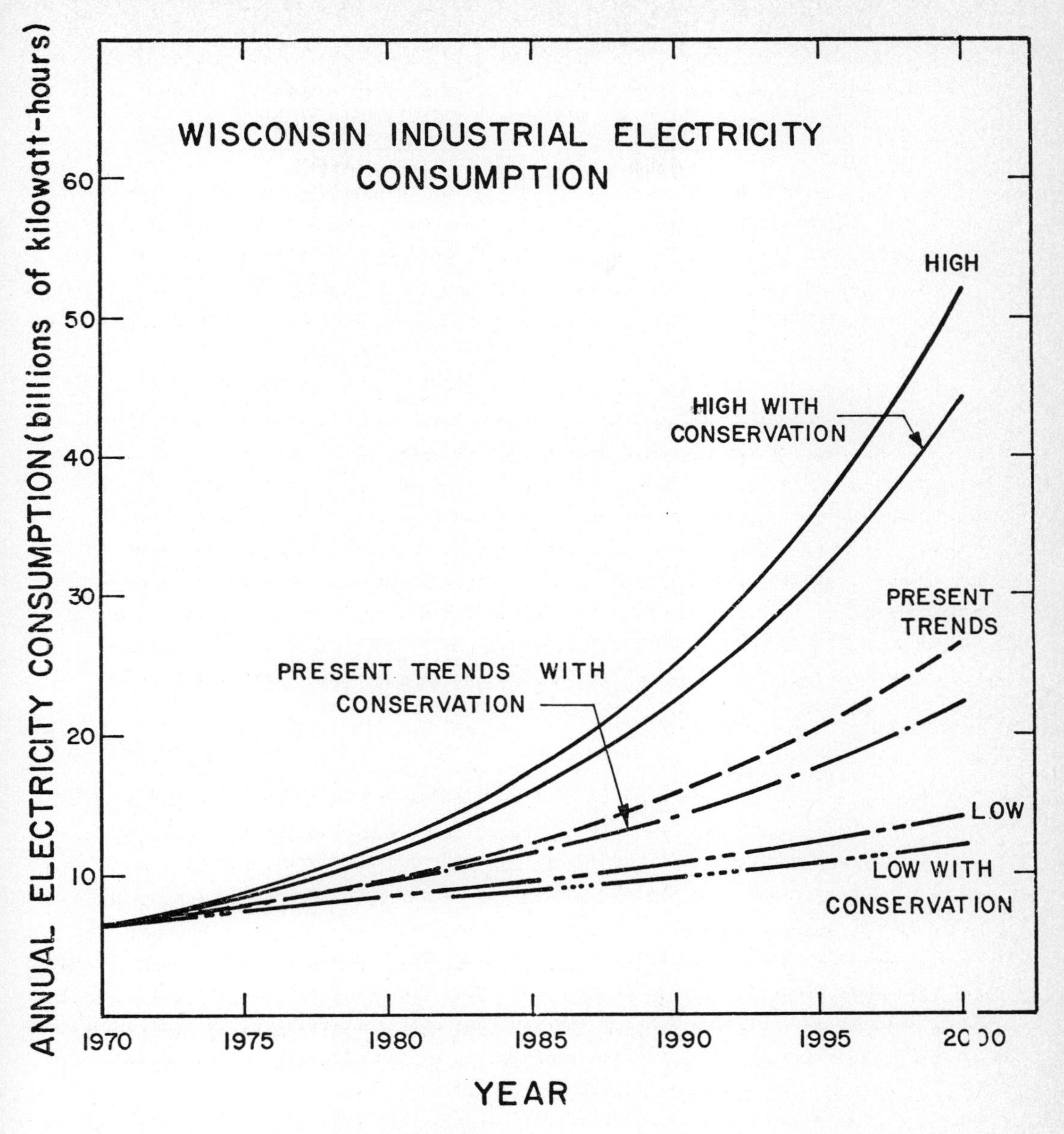
WISCONSIN INDUSTRIAL ELECTRICITY CONSUMPTION
ANNUAL ELECTRICITY CONSUMPTION (billions of kilowatt-hours)
60
50
40
30
20
10
HIGH
HIGH WITH CONSERVATION
PRESENT TRENDS
PRESENT TRENDS WITH CONSERVATION
LOW
LOW WITH CONSERVATION
1970
1975
1980
1985
1990
1995
2000
YEAR

FIGURE 11

Table 4

ASSUMED ANNUAL GROWTH RATES IN PER CAPITA COMMERCIAL SALES AND GOVERNMENT EXPENDITURES

Case Numbers	Growth Rates (% per year)	
	Commercial Sales Per Capita	Government Expenditures Per Capita
1 and 4	2.3	3.0
2 and 5	2.5	3.5
3 and 6	0.0 -by 1990	0.0 -by 1990

Source: Energy Systems and Policy Research Group

Table 5

ASSUMED ANNUAL GROWTH RATES IN ELECTRICITY INTENSIVENESS FOR THE COMMERCIAL SUBMODEL

Case Numbers	Electricity Intensiveness Growth Rate (% per year)
1 and 4	varies from 5.0 in 1970 to 2.0 in 1980, 2.0 for 1980-2000
2 and 5	varies from 5.0 in 1970 to 3.5 in 1980, 3.0 in 1990, 2.0 in 2000
3 and 6	Same as Present Trends

Source: Energy Systems and Policy Research Group

The alternative electricity consumption projections are displayed in Figure 12. The wide range of demands in the year 2000, from 77.0 billion KWH for the High Case, down to 19.9 billion KWH for the Low Case with conservation, is indicative of the broad range of possibilities represented by the alternative assumptions.

IV.1.4 Total Requirements for Electricity Generation: The electricity consumptions given above for the three sectors all represent end-use electricity. In order to calculate total electrical generation, it is necessary to determine transmission and distribution losses. Losses for the country as a whole, as well as for Wisconsin, averaged 9.5% of the total generation for the period 1970-1972. Therefore, it has been assumed for Cases 1-6 that 9.5% of the total generation is lost before the electricity arrives at the end user.

Total electrical generation in Wisconsin for Cases 1-6 shown in Figure 13

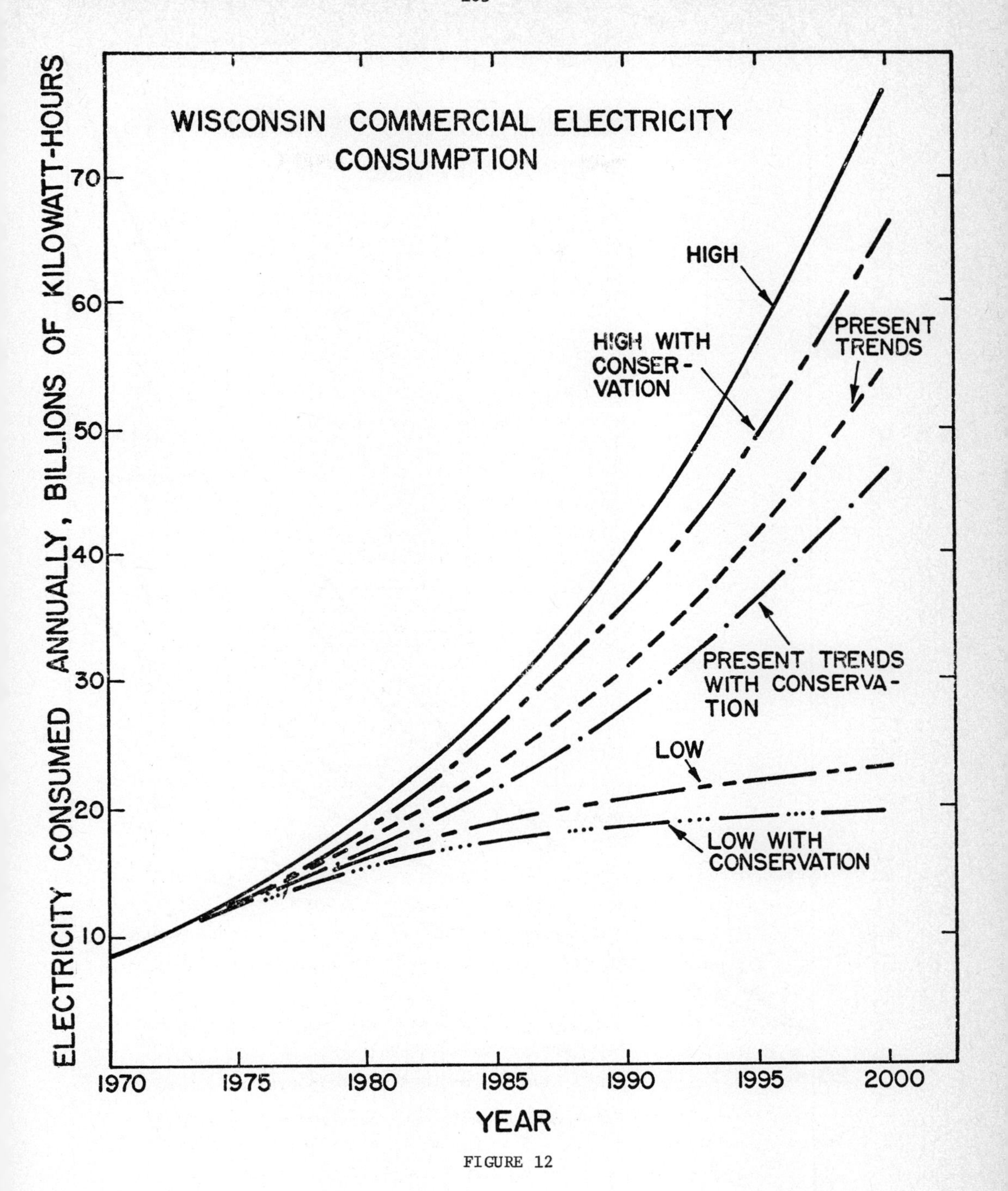
WISCONSIN COMMERCIAL ELECTRICITY CONSUMPTION
ELECTRICITY CONSUMED ANNUALLY, BILLIONS OF KILOWATT-HOURS
70
60
50
40
30
20
10
1970
1975
1980
1985
1990
1995
2000
YEAR
HIGH
HIGH WITH CONSER-VATION
PRESENT TRENDS
PRESENT TRENDS WITH CONSERVA-TION
LOW
LOW WITH CONSERVATION

FIGURE 12

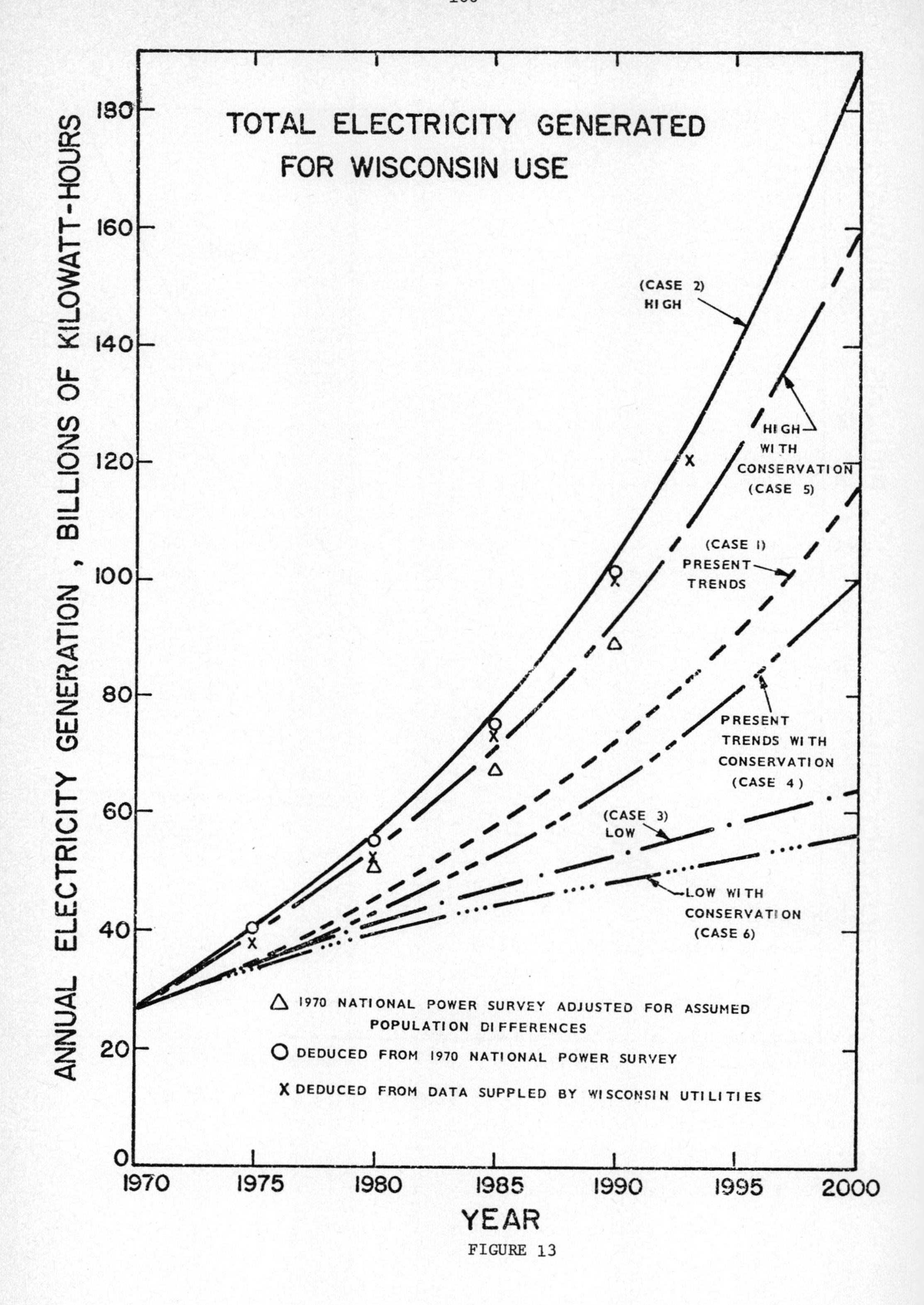
TOTAL ELECTRICITY GENERATED
FOR WISCONSIN USE
ANNUAL ELECTRICITY GENERATION, BILLIONS OF KILOWATT-HOURS
0
20
40
60
80
100
120
140
160
180
1970
1975
1980
1985
1990
1995
2000
YEAR
(CASE 2)
HIGH
HIGH
WITH
CONSERVATION
(CASE 5)
(CASE 1)
PRESENT
TRENDS
PRESENT
TRENDS WITH
CONSERVATION
(CASE 4)
(CASE 3)
LOW
LOW WITH
CONSERVATION
(CASE 6)
△ 1970 NATIONAL POWER SURVEY ADJUSTED FOR ASSUMED POPULATION DIFFERENCES
○ DEDUCED FROM 1970 NATIONAL POWER SURVEY
X DEDUCED FROM DATA SUPPLED BY WISCONSIN UTILITIES

FIGURE 13

is the summation of demands from the residential, industrial, and commercial sectors plus the losses. It is interesting to note the extreme divergence of the cases. By 1985, the High Case (2) has nearly 75% more electrical generation than the lowest case (6). By the year 2000 the difference between those two cases is more than a factor of three. The Present Trends Case (1) falls about in the middle of the range shown. Also shown in Figure 13 are generation estimates for Wisconsin deduced from the 1970 National Power Survey, (14) and from demand and generation information presented by the Wisconsin utilities at the Energy Hearings held by the Wisconsin Public Service Commission (PSC) in July 1973 (15). As Figure 13 indicates, the FPC-Wisconsin Utilities results are close to the High Case (2).

Although accurate short-term forecasting is not the primary objective of the WISE model, it is, nevertheless, interesting to compare the actual generation in Wisconsin for the years 1970 to 1971 with the alternative futures obtained from the Wisconsin Regional Energy Model. The comparison is shown in Figure 14, in addition to the forecasts by the Federal Power Commission and the Wisconsin utilities. It should be noted that the computer simulation with the WISE Model began with the year 1970 and contained no historical data or statistics for the years following 1970. The computed results in 1971-1973 were therefore completely independent from the actual generation in those years. Data for the first six months of 1974 indicate that electricity generation by the largest utilities in Wisconsin is only about one percent above the generation in 1973. The average annual increase in Wisconsin's generation from 1970 to 1973 was 5.6 percent. For the period 1960 to 1970, an annual growth rate somewhat greater than seven percent was typical.

These forecasts are described in detail in a recent report by the Energy Systems and Policy Research Group (12). They have also been presented in formal testimony at Wisconsin Public Service Commission hearings, and are being studied by the Commission in their evaluation of future generation requirements.

None of these electricity futures should be considered as the "best" forecast but rather as providing an indication of the response of the Wisconsin energy system to changes in some of the factors responsible for electricity demands and as drawing some boundaries on what the future may hold. They also provide a strong basis to analyze and quantify many of the building blocks of the future. Perhaps most importantly, they serve to demonstrate the very broad spectrum of energy futures open to Wisconsin -- and the continuing need for a more explicit discussion of the decisions leading to those futures.

IV.2 Energy Conservation Measures in Wisconsin

Wisconsin is an energy consuming state, located at "the end of the pipeline"

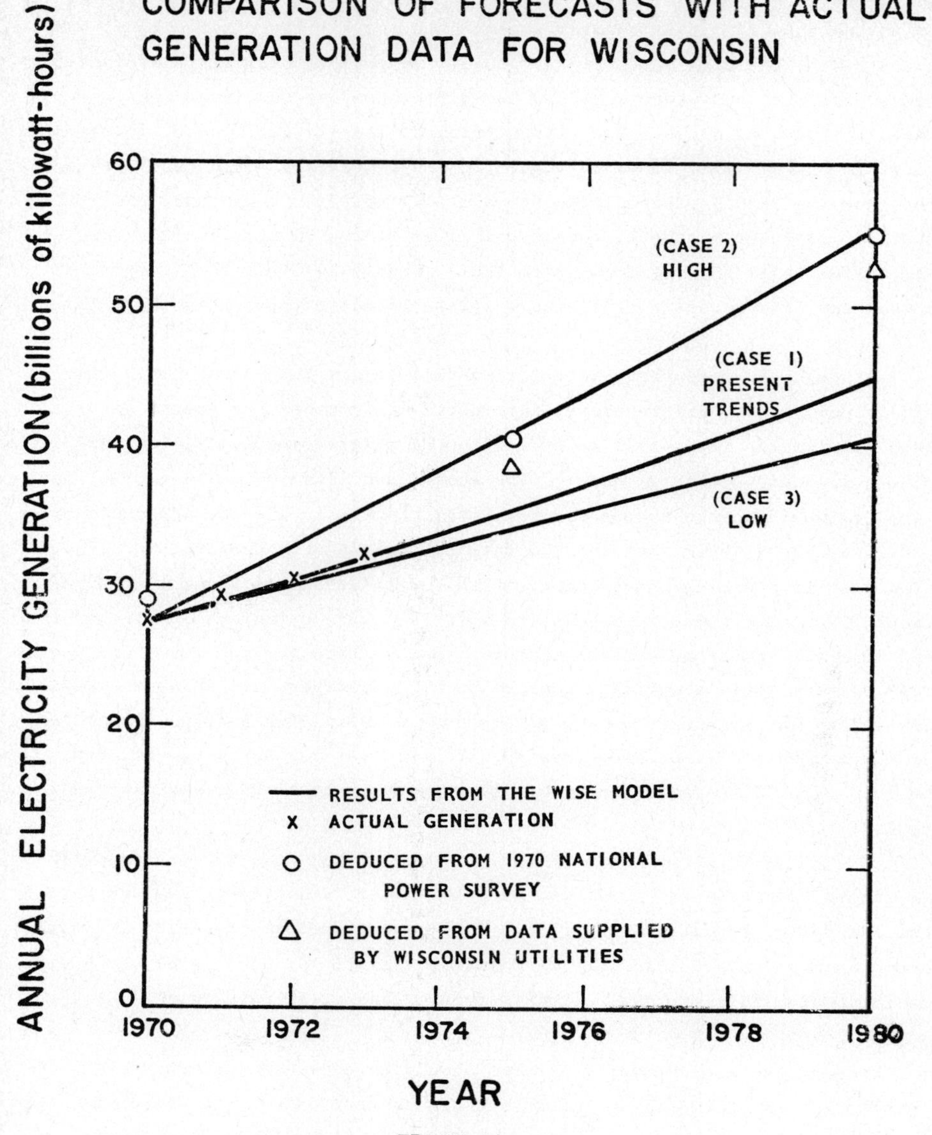
COMPARISON OF FORECASTS WITH ACTUAL GENERATION DATA FOR WISCONSIN
ANNUAL ELECTRICITY GENERATION (billions of kilowatt-hours)
60
50
40
30
20
10
0
1970
1972
1974
1976
1978
1980
YEAR
(CASE 2) HIGH
(CASE 1) PRESENT TRENDS
(CASE 3) LOW
RESULTS FROM THE WISE MODEL
X ACTUAL GENERATION
DEDUCED FROM 1970 NATIONAL POWER SURVEY
DEDUCED FROM DATA SUPPLIED BY WISCONSIN UTILITIES

FIGURE 14

with no fuel resource of its own. As protection against both increasing security and cost of all forms of energy, the people of Wisconsin have a special interest in ensuring that they follow prudent energy management practices. This is particularly important in the case of construction of semi-permanent energy-consuming facilities such as buildings, transportation systems, durable goods, etc.

In response to this concern, the Wisconsin State Legislature has created an Energy Conservation Advisory Council, comprised of members of key state agencies, legislators, and several representatives from the public at large. The Council has devoted the past eight months to examining the feasibility and potential impact of a wide range of energy conservation measures in Wisconsin.

In a related but narrower action along these same lines, an Energy Conservation Committee has been appointed to develop improved energy management specifications in the statewide commercial building codes administered by the Wisconsin Department of Industry, Labor and Human Relations (DILHR). This committee, composed of technical experts from several disciplines, is advising DILHN on short- and long-range changes in the Wisconsin Administrative Building Code which will aid in energy conservation. The Energy Systems and Policy Research Group has been an active participant in both of these State energy conservation activities, in part through formal membership of individuals on the committees, but also as a source of technical expertise and recommendations.

In connection with these and other conservation activities, the WISE Model has been applied to the analysis of several alternative Wisconsin energy futures based upon the implementation of improved energy management measures. A selected few of these are described below.

IV.2.1 Use of More Efficient Air Conditioners: A significant reduction in electricity needed for room air conditioners in the State of Wisconsin could result from the use of more efficient air conditioners. The residential submodel of WISE was applied to the study of the potential energy-use reductions in three of the previously described electricity scenarios, namely, the Present Trends, High, and Low Cases.

The potential savings were based upon the postulated introduction of air conditioners that require 40 percent less electricity than the present average to provide the same amount of cooling capacity. Units are currently available that use 50 percent less than the average unit (16). The average life of an air conditioner was assumed to be 10 years. The improved efficiency unit is assumed to be exclusively purchased starting in 1976 for the conservation cases (4-6), while in the cases without conservation (1-3), the efficiency of the average unit was assumed to remain the same as it had been. Based upon an average power rating of 1.566 kilowatts and the assumption that the air conditioners all operate the same 575 hours per year in Wisconsin, the following reduction in summer peak loads is obtained with improved efficiency air conditioners.

	1985	2000
Present Trends (1 and 3)	687 MWE	967 MWE
High Cases (2 and 5)	965	1202
Low Cases (3 and 6)	687	967

The State of Wisconsin is now seriously considering proposals to achieve the above savings by establishing implementation of minimum efficiency-standards and/or efficiency-labeling requirements for air conditioners.

IV.2.2 A Shift Toward Smaller Passenger Autos

Wisconsin is also considering a number of policy alternatives aimed at reducing or slowing the growth of energy use of its transportation systems. Among these are various measures to provide incentives for a shift toward the use of smaller and more energy-efficient passenger autos. To evaluate the potential energy savings from various measures, the transportation submodel of WISE was used to examine several alternative energy futures through the year 2000. The following fuel efficiencies were used:

	Energy Efficiencies (Miles per Gallon)	
	Intra-city	Inter-city
Conventional cars	10	13
Small cars	20	22

The two cases examined were identical except for the assumption that in Case I, 20% of the automobiles entering the vehicle population each year are small cars and 80% are conventional cars. Case II, 20% and 80% are used as starting values (1970) but the percentage of small new cars increases by 2.67% each year until 1985 when almost all new cars are small and 40% are conventional. The rate of increase of small cars is then decreased to 1.34% increase each year. By the year 2000 under this scenario, the percentages of conventional and small cars have been reversed with 20% of the new passenger vehicles each year being conventional automobiles and 80% small automobiles.

The savings resulting from this shift are significant, as shown in Figure 15. In the year 1980, the savings is 85 million gallons of gasoline or 6 1/2% of the total used for passenger vehicles in that year. By the year 2000, the savings is 488 million gallons or 27% of the total. Cumulative savings from 1970 to the year 2000 is 5940 million gallons. A number of other energy futures, including shifts to mass transit and introduction of the electric auto have also been examined

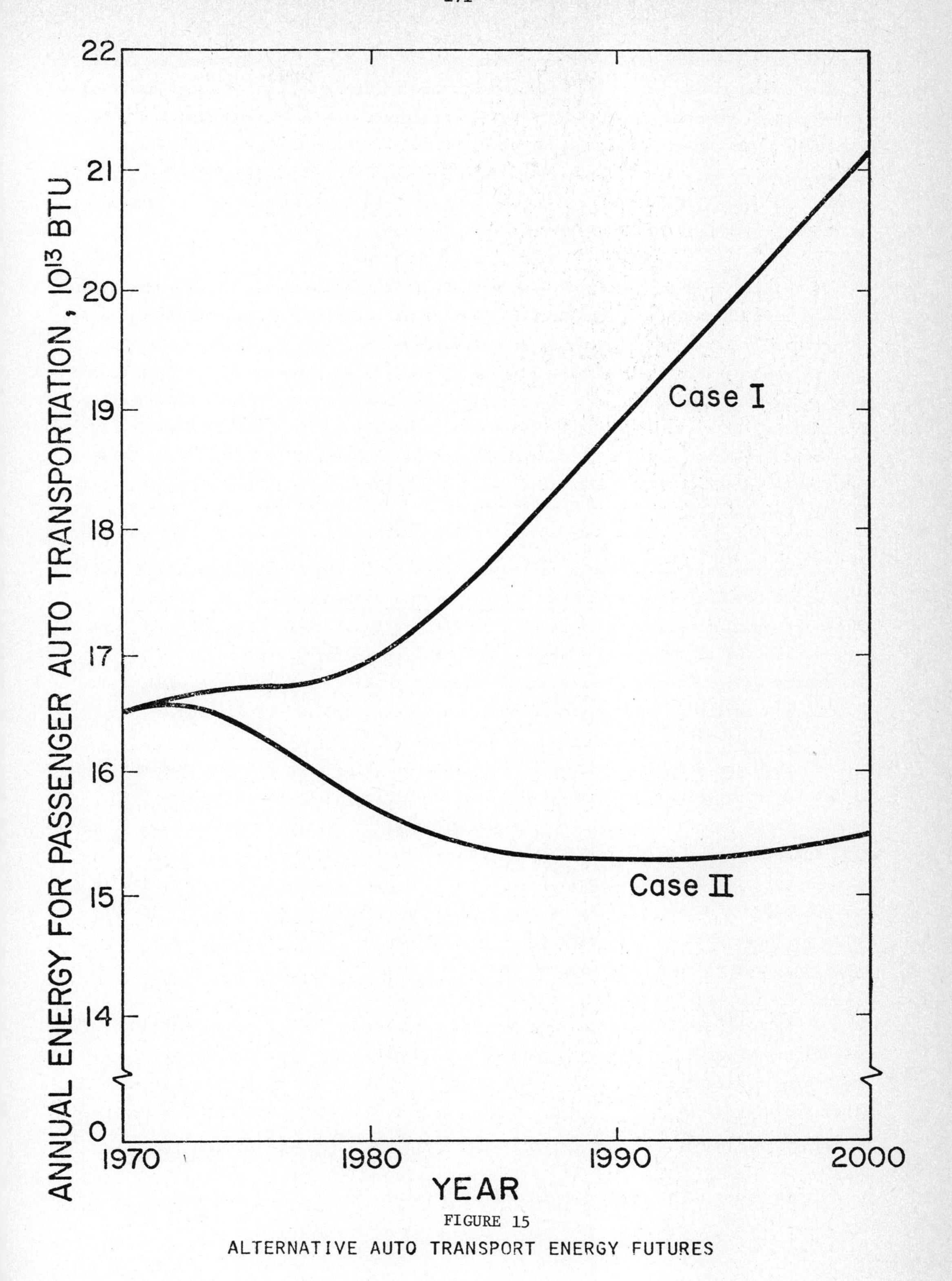

FIGURE 15

ALTERNATIVE AUTO TRANSPORT ENERGY FUTURES

IV.2.3 Improved Building Insulation Practices. The State of Wisconsin is currently holding hearings to receive public comments on its recently announced proposals for conservation-oriented changes in the state building code for commercial and public buildings. In addition, the Energy Conservation Advisory Council is developing legislation to establish statewide energy-conserving building codes for single family residences. The WISE Model has been employed to examine various futures associated with both of these proposed regulations.

Figure 16 illustrated two possible futures associated with differing degrees of home insulation. Case I assumes all homes are insulated at the "Minimum Property Standards" level established by the Federal Housing Administration in 1970. Case II assumes that insulation would yield "maximum economic benefit" (16) to the home owner is installed in all new homes. Other assumptions include the continued availability of natural gasoline residential customers and a continuation of old home demolition and renewal at approximately the 1970 rate. Oil, coal and other fuels were included in the calculations although they are not discussed explicitly in this section.

By the year 2000, there is a 26% reduction in the energy used for residential space heating. Based upon an average annual energy savings of 2.89×10^{13} BTU, the total cumulative savings from 1970 to 2000 is equal to 86.61×10^{13} BTU, approximately 70% of the total energy used in Wisconsin for all purposes in 1970. The energy savings would be even greater if air conditioners were included, i.e., in addition to the savings would be gained from the improved insulation.

Both the proposed Wisconsin commercial building code and the proposed ASHRAE building guidelines have been examined within the framework of the WISE Model. These studies were based upon the engineering-design version of the commercial submodel and a further refinement (17) which includes a more detailed treatment of building design characteristics such as the average number of stories in new buildings and the fraction of the exterior area devoted to windows. The results of these studies have been quite valuable in assessing the statewide savings resulting from various forms of the code.

IV.2.4 Natural Gas Diversion. Due to a growing concern about the possibility of natural gas shortfalls, the Wisconsin Office of Emergency Energy Assistance asked the Energy Systems and Policy Research Group to analyze the impact on Wisconsin of diversion of various quantities of this primary fuel. The WISE Model was employed in this analysis and among the results generated were those described below.

In order to estimate the impact of a diversion, several assumptions must

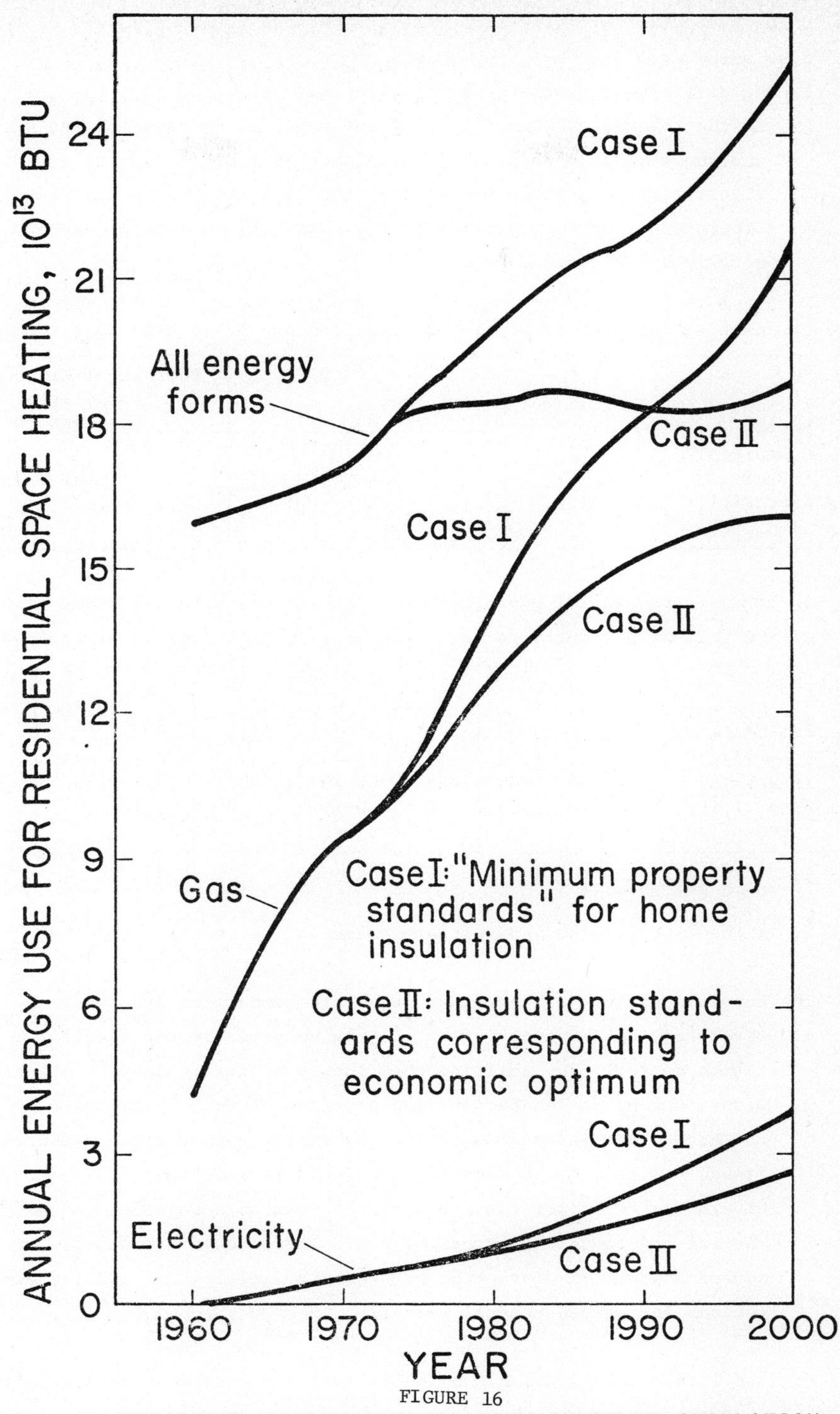

FIGURE 16

WISCONSIN ENERGY SAVINGS FROM IMPROVED INSULATION

be made. For our analysis of the impact of a diversion spread out evenly over the year 1975 we assumed:

1. No diversion from the residential sector
2. In the commercial sector, 12% of the usage is by hospital and schools, and there will be no diversions in those buildings
3. Electricity generation use will be the first use to be curtailed
4. Diversions from the commercial and industrial sectors are spread equally by percent

The natural gas uses from which a diversion may be obtained are then:

Sector	10^{12} BTU	Percent of Wisconsin Yearly Total Consumption
Commercial	41	11
Industrial	154	39
Electricity Generation	28	7
	223	57

An analysis of natural gas consumption in sectors with allowable diversion shows that the following reductions will be necessary with various diversions:

Sector	Percent Diversion 2% E[1]	%[2]	10% E	%	20% E	%	25% E	%
Commercial	41	100	39	95	30	74	20	50
Industrial	154	100	145	95	115	74	77	50
Electricity Generation	20	71	0	0	0	0	0	0

[1] E is in 10^{12} BTU/yr

[2] % is the % of original, divertable usage in the sector.

The conclusions that can be drawn from this comparison are that:

1. Diversions up to seven percent can be absorbed by the electrical generation sector. The shift to coal should be fairly easy by utilities.
2. Diversions up to ten percent can probably be readily absorbed by the commercial and industrial sectors by energy conservation measures. Probably little or no substitution would be required.
3. Diversions of 25 percent would reduce available supplies in the commercial and industrial areas by 50%. There would undoubtedly be severe short term disruption of economic activity and massive substitutions and shifts to coal and oil.

We have little data available on short-run fuel substitution possibilities. The previous estimates have been based on the assumption of no fuel substitutability.

For complete substitutions we can estimate the additional tons of coal or fuel oil required as:

Percent Diversion	Coal (10^6 tons)	Fuel Oil (10^6 bbl)
2	0.3	1
10	1.3	6
20	2.6	11
25	3.3	14

V REMARKS

This report has presented a general description of the research efforts of the Energy Systems and Policy Research Group at the University of Wisconsin-Madison. The energy model described herein is dynamic not only in structure but also in form as it is continuously modified in response to feedback from its users and its audience. The model has proven useful and usable for analysis of a wide variety of energy related issues and it is our hope that the current modifications and improvements will enhance its usefulness in these endeavors.

N.B. The reference number in parentheses used by the authors in their text are given at the end of each reference.

REFERENCES

Buehring, J.S. and P.D. Kishline, 1974. "WISSIM - A Simulation Command Language, Part 2: User's Guide for Interfacing WISSIM and Simulation Models," Energy Systems and Policy Research Memo #13, Institute for Environmental Studies, University of Wisconsin-Madison, November. (11)

Buehring, W.A., et al., 1974. "Alternative Future Electricity Generation Requirements for the State of Wisconsin," Energy Systems and Policy Report #4, Institute for Environmental Studies, University of Wisconsin-Madison, October. (12)

Buehring, W.A. and W.K. Foell, 1974. "A Model of the Environmental Impact of Wisconsin Electricity Use," Energy Systems and Policy Research Report #6, Institute for Environmental Studies, University of Wisconsin-Madison, November. (9)

Buehring, W.A., W.K. Foell, and R.L. Dennis, 1974. "Environmental Impact of Regional Energy Use: A Unified Systems Approach," Energy Systems and Policy Research Group, Institute for Environmental Studies, University of Wisconsin-Madison, September. (8)

Dietz, M.M. and W.K. Foell, 1974. "1974 Survey of Energy Use in Wisconsin," Energy Systems and Policy Research Report #3 and Institute for Environmental Studies Report #25, University of Wisconsin-Madison, October. (2)

Executive Office of the President, 1972. Office of Emergency Preparedness, The Potential for Energy Conservation, A Staff Study, (OEP 4102-0009), October. (13)

Frey, D.A., 1974. "A Model of Residential Energy Use in Wisconsin," Energy Systems and Policy Report #11, Institute for Environmental Studies, University of Wisconsin-Madison, November. (4)

Hirst, E. and J.C. Moyers, 1973. "Efficiency of Energy Use in the United States," Science, March 30. (16)

Jacobson, D.A., J.W. Mitchell, and J.L. Pappas, 1974. "A Model of Commercial Energy Use in Wisconsin," Energy Systems and Policy Report #5, Institute for Environmental Studies, University of Wisconsin-Madison, November. (5)

Kishline, P.D. and J.S. Buehring, 1974. "WISSIM - A Simulation Command Language, Part I: User's Guide to WISSIM Commands," Energy Systems and Policy Research Report #9, Institute for Environmental Studies, University of Wisconsin-Madison, November. (10)

Mitchell, J.W. and M.A. Caruso, 1975. "A Model of Transportation Energy Use in Wisconsin," Energy Systems and Policy Research Report #8, Institute for Environmental Studies, University of Wisconsin-Madison, March. (1)

Mitchell, J.W. and D. Jacobson, 1974. "Implications of Commercial Building Codes for Energy Conservation," Energy Systems and Policy Research Report #15, Institute for Environmental Studies, University of Wisconsin-Madison, December. (17)

Peerenboom, J.P., 1974. "A Simulation Model for Examining Long-Range Electric Generating Capacity Expansion Alternatives for Wisconsin," Unpublished M.S. Thesis, Department of Nuclear Engineering, University of Wisconsin-Madison. (6)

Peerenboom, J.P., W.K. Foell, and J.L. Pappas, 1974. "A Model for Examining Long-Range Generating Capacity Expansion Alternatives for Wisconsin," Energy Systems and Policy Research Report #12, Institute for Environmental Studies, University of Wisconsin-Madison, November. (7)

Reports Submitted by Wisconsin Electric Utilities to the Public Service Commission of Wisconsin Pursuant to Phase I of Order 2-u-7643, July 1973. (15)

Shaver, D.B., J.L. Pappas, and W.K. Foell, 1975. "The Wisconsin Industrial Energy Use Model: Description of the Model and Analysis of Alternative Futures," Energy Systems and Policy Report #14, Institute for Environmental Studies, University of Wisconsin-Madison, March. (3)

U.S. Federal Power Commission, 1970. The 1970 National Power Survey, U.S. Government Printing Office. (14)

ESTIMATING THE REGIONAL IMPACTS OF ENERGY SHORTAGES

William Donnelly and Ali M. Parhizgari[1]

I INTRODUCTION

The current state of regional modeling makes it extremely difficult to evaluate comprehensively (1) the impact of shocks to a regional economy or (2) the regional implications of different national policies. Theory provides basic insights into the workings of a region's economy, but the framework of a general model is usually not convenient for analyzing specific policy questions. Much regional analysis is currently directed to the answering of energy policy questions (Cumberland, 1975 and Nisson et.al. 1975).

Improvement of the situation in regard to specific questions on changes in regional economies may come about in several ways: (1) by the utilisation of experts' opinions, (2) by adapting an existing regional model to answer policy questions. Although valuable the first option provides variable results, since it relies upon subjective--albeit educated--opinions. Any biases are implicit; and therefore, the conclusions rendered by such an approach are difficult to evaluate and criticize. In spite of this, the "Delphi" technique a form of judgemental modeling, is much in vogue, and seem to be gaining wide acceptance today.

The second option is rejected because of the costs, in terms of person-years and financial expense, involved in constructing a comprehensive regional model designed for evaluating policy issues. Constructing an energy-specific, regional macro-economic model and its concomitant data base is not considered feasible for a project with a short planning horizon.

The third option, adapting and applying an existing model, is judged a more reasonable approach, at least for immediate purposes. It eliminates duplication of an effort in the preparation of the requisite regional data base, and it saves time and resources. With appropriate modifications, the final results should be similar to those from a model actually developed from scratch.

---

[1]This research was supported under a grant from the Economic Development Administration. The opinions expressed are the authors and do not necessarily represent the views of the Economic Development Administration, the Federal Energy Administration or the Maryland Department of Economic and Community Development. We would like to thank Lloyd Atkinson and Curtis Harris for their comments and suggestions. All errors which remain are the authors alone.

Several existing regional models have been reviewed and analyzed in terms of their adaptability and theoretical validity. The authors conclude that the Maryland Multiregional, Multi-Industry Forecasting Model (Harris, 1973) henceforth referred to as the regional model, offers the most versatility of any of the existing regional models. The following discussion of a particular shortage analyzed using this model supports this contention and offers suggestions for further improvements and extensions to the model.

II NATURAL GAS CURTAILMENT

The recent national fuel shortages have stimulated considerable interest in forecasting the expected impacts of these shortfalls on local economies. In this context, the regional model is utilized to estimate the economic impacts on the State of Maryland due to curtailment of natural gas deliveries. Two alternative allocation schemes for distributing the anticipated reduction of natural gas supply are evaluated. First, the impacts of the priority end-use plan advocated by the Federal Power Commission (FPC) are considered. This plan, which ranks customers on an ascending scale from one to nine by end-use category, requires that interstate pipelines allocate their available supply of natural gas to distributors according to the end-use of the gas. This plan is outlined in FPC Order No. 467-C, April 4, 1974, pp. 3-5. Conversely, the State of Maryland and the Columbia Gas Transmission Corporation (Columbia) advocate continuing the current allocation system which curtails only interruptible customers and distributes proportionately any further reductions across all classes of natural gas consumers. Interruptible natural gas service was originally conceived and established as a "buffer for possible expansion in the residential-commercial sector" (Balestra, 1966) and is therefore designed to alleviate excess capacity due to the capitalization required in the expansion process and to mitigate peak load problems.

The pro rata position results in, initially at least, a higher allocation of natural gas to regions having larger relative proportions of large industrial users. The short-run trade-off is essentially between residential and industrial usage. The pro rata plan is intended to allow time for the conversion of plant and facilities to alternative fuels. The argument is that a gradual transition from the existing situation would result in less severe impacts on the economy than the abrupt suspension of service that is required under the end-use plan.

Basically, therefore the two allocation procedures differ as to the absolute amounts of natural gas to be delivered to the different service areas. The question is how the anticipated reduction in supply is to be shared among customers and among the regions served. Generally, the market mechanism is considered a most efficient rationing device. However, when an excess demand for a commodity exists at the regulated price, the price system will not function to allocate the item appropriately. Since either the immediate deregulation of the natural gas industry or the abrupt curtailment of deliveries of natural gas to large industrial consumers would result in severe economic repercussions, some transitional alternative should be considered.

Table 1 presents the estimated curtailment to the distributors served by Columbia over their multi-state service area under the recommended pro rata plan. It can be seen that the pro rata plan would result in an 11.6 percent reduction in deliveries beginning with the winter season (1974-75) (the natural gas year, as defined by the industry, extends from November through October). This projected reduction would gradually increase to a 22.6 percent figure by 1977-78.

Table 1

Projected Pro Rata Curtailment of Natural Gas Supplies
(Columbia's Multi-state Distribution Area)

Year	Projected Percentage Curtailment (1973-74 Base Year)
1974-75	11.6
1975-76	13.7
1976-77	17.1
1977-78	22.6

Conversely, the immediate impact of the FPC proposed priority end-use allocation scheme would be a reduction in supply of natural gas to the State of Maryland by 20.6 percent. Table 2 shows the projected percent curtailment in the State for four natural gas seasons under the end-use plan.

Table 2

Proposed Priority Plan Curtailment in Natural Gas
Deliveries (State of Maryland)

Year	Projected Percentage Curtailment (1973-74 Base Year)
1974-75	20.6
1975-76	21.4
1976-77	22.9
1977-78	24.9

Thus under the FPC plan the total impact is immediate and remains relatively constant over the entire period. The direct burden of the natural gas shortage falls primarily on the industrial sectors, because residential and small commercial users are assigned the highest priorities. In addition, areas having a relatively high level of large industrial customers will experience under the end-use plan a higher initial cut-back in natural gas supplies than will occur in primarily residential or commercial service areas. Thus the more industrialized service areas will suffer larger direct reductions in natural gas, at least in the short-run.

III MARYLAND MULTIREGIONAL MULTI-INDUSTRY FORECASTING MODEL

The regional model forecasts changes in the location of industrial production. It is a spatial allocation model that distributes changes in industry output as a function of economic variables. The primary variables used in the model are: (1) location rents, (2) the value of land, (3) the existence of capital stock, (4) the location of demand for outputs and (5) the availability of supply of inputs.

The model utilizes the county as the basic unit of spatial disaggregation. The 3,111 counties, or county equivalent areas, in the United States represent a set of mutually exclusive, exhaustive regions into which the county may be divided. The counties are not envisioned as representing anything more than political units for which regular and consistent data series are available. This choice of data set enables the regional modeler to work with large varieties of regions, since any set of regions, composed of county units, can be utilized. Therefore the definition of the regions depends upon the problem being analyzed. Thus the counties may be used as spatial building blocks to construct any set of larger regions, such as Functional Economic Areas. Functional Economic Areas are relatively closed or bounded areas with respect to income-reproducing and consumer-oriented or "residentiary" activities (Fox and Kumar, 1965). While a county level data set offers great flexibility in defining regions, it also represents a major disadvantage because of the compilation and computational demands imposed. The ambitious nature of the Harris project is illustrated by the choice of the county as the basic spatial building block.

The ninety-nine industry sectors of the model are defined in terms of the Standard Industrial Classification (SIC) code. Each of the industry sectors has a separate equation which explains the location of changes in output. This is accomplished by using two sets of variables: (1) input prices that firms face in each location and (2) agglomeration variables that help explain location behavior that is not accounted for by prices. The agglomeration variables relate to: (1) the skill level of the labor force, (2) the quality of the transportation system, (3) the level of the existing infra-structure, or (4) the variety of potential suppliers and consumers for the product. This degree of industry detail facilitates the investigation of changes in the industrial structure of an area. Such detail also imposes substantial costs in terms of data requirements.

The data base requirements for implementing the regional model are impressive. Direct data are available for very few of the variables incorporated in the model. Regional estimates of personal consumption expenditures by industry sector are obtained from the Bureau of Economic Analysis, U.S. Department of Commerce. Similarly, estimates of defense expenditures and gross foreign exports are also available from Federal government sources. Estimates of regional sector specific employment are derived from County Business Patterns. The remaining regional data required bythe model are constructed by applying national coefficients. Given the absence of regional data, this is a reasonable alternative.

Such data exigencies are acknowledged realities in regional modeling work. All regional analysis suffers from the paucity of good, consistent data. This problem arises equally for the simpler descriptive techniques as well as for the more theoretically elegant and data-intensive econometric procedures. Even the most naive procedures, i.e., trend extrapolation, economic base studies or shift-share analysis, require large amounts of consistent regional data which are not generally available, at least not in a readily usable form. The general lack of suitable regional data has lead to the development of many techniques designed to adjust national figures to regional levels (Polenske, 1970). Since the regional values for many variables are not reported, there exists little basis for comparing and evaluating the quality of the various approximating techniques developed and utilized. Limited analysis has been attempted for decenniel Census years, however. The conclusion is that careful scrutiny must be given to any regional variables constructed from national figures (Brown, 1969). Thus judgements on the quality of regional research must distinguish between questions of the theoretical structure of a particular model and the data exigencies which all regional models face.

IV ESTIMATES AND FORECASTS

The regional model utilizes estimates according to the theoretical relationships discussed in the literature. Essentially, the location of new production capacity is determined so as to minimize the marginal transport costs of both commodity inputs and outputs given the location of the existing productive capacity (Harris and Hopkins, 1972).

The forecasts generated are recursive in nature with one year's output serving as the input data for subsequent years. The actual mechanics are as follows:

(1) estimate the supply and demand for a particular industry at each location for the base year and the year prior to it;

(2) estimate shipping costs of a unit of the industry's output;

(3) derive the shadow prices of shipping an additional unit of all primary inputs and outputs from a linear programming transportation model;

(4) estimate the parameters in the industry location equations;

(5) enter the base year values and generate regional forecasts of the change in industry output;

(6) forecast capital expenditures;

(7) forecast regional employment and payrolls and adjust forecasts from an "establishment" basis to a "residence" basis;

(8) forecast regional population and migration and calculate labor force;

(9) derive estimates of payrolls and other components of personal income and forecast consumption expenditures;

(10) update exports, imports and government expenditures and generate forecasts of intermediate demand;

(11) return to step (3),or to step (5).

Thus the model is not of the input-output genre. Regional input-output models begin with a set of fixed technical production coefficients. The coefficients in input-output tables are generally held invariant with respect to relative price changes, technological changes, or trade flow changes. In such models, regional final demand is predetermined with the input-output table used to generate estimates of the intermediate flows. The most comprehensive regional input-output work to date is under the direction of Karen Polenske (Polenske, 1970).

The simplifying assumption of fixed coefficients may be acceptable in the short run for a national input-output table, since a nation such as the United States represents a relatively closed system and the aggregate relationships change rather slowly over time. In the preparation of regional forecasts this approach is less justifiable because the assumptions of fixed technical and trade coefficients are too restrictive when considered in the context of a small, open economy (Conway, 1975). Conversely, the Harris approach incorporates regional behavioral equations and estimates of the marginal transportation costs in a spatial allocation model. Moreover this form of spatial allocation model does not require estimation of the inter-regional trade coefficients which may be expected to change from period to period.

The operation of the regional model requires a set of internally consistent national forecasts. These national forecasts are used as controls to assure that reasonable forecasts are produced. In this context "reasonable" should be interpreted as nationally achievable and consistent values. The actual procedure used to generate the national values is immaterial to the operation of the regional model. Currently, the model relies upon the national forecasts derived from a dynamic input-output model (Almon, 1974). The Interindustry Forecasting Model of the University of Maryland (INFORUM), incorporates estimates of technological change as well as equations which forecast changes in prices and commodity substitution. These later effects are included to allow for product substitution that occurs as a result of changes in relative prices. This is a means of overcoming the fixed coefficient assumption of traditional input-output tables.

Any set of consistent national forecasts could be used for the balancing process in the regional model regardless of their type; e.g., forecasts derived from a macro-economic model with sectoral disaggregation, would be equally suitable. The INFORUM input-output coefficients also serve to calculate estimates of the regional intermediate demand figures in Step 10 mentioned before. Finally, the INFORUM model offers considerable industry detail that is not available in most other existing econometric models, while providing input-output coefficients that change over time.

V IMPACT TESTS

The regional model is utilized to determine the relative impact on the State of Maryland of two allocation plans, i.e., a priority end-use scheme and a pro rata plan. The direct, indirect, and induced effects of these two allocation plans are evaluated in terms of their impacts in terms of changes in output, employment, earnings, personal consumption expenditures, and investment.

To start with, an initial run is made under the assumption of no shortages of natural gas. This serves as the benchmark against which the two allocation plans are measured. The second task is to incorporate the two allocation plans into the model. The procedure chosen to incorporate the anticipated shortages is essentially "data-oriented" and is briefly described below.

Baltimore Gas and Electric Company and Washington Gas Light Company provided a list of their customers on interruptible service ranked according to the FPC priority categories. The Maryland Department of Employment and Social Services provided the average monthly employment figures for these firms. Estimates of the total amount of curtailment under the two plans were made available by Columbia. The following steps are followed for each of the allocation plans:

(1) Determine the number of days of curtailed service to interruptible customers under both allocation plans

(2) Aggregate (or disaggregate) the "firm" employment data into the industrial sectors and counties (Table 3, Col. 5)

(3) Calculate the percentage of the total sector employment

(3) (continued)
that is represented by the firms on interruptible services (Col. 5 divided by Col. 4, Table 3)[2]

(4) Adjust the estimates of curtailment, by the percentages obtained in Step 3. The resulting figures give the percentages by which the output of each sector in each county is initially reduced.

(5) Reduce a sector's output for the counties affected by the percentages obtained in Step 4 and re-run the model using these adjusted outputs.[3]

(6) Compare the resulting employment, output, and other indicators with the benchmark values (see Tables 4 and 5).

The model is run to determine the total impact implied by the direct reduction in the output resulting under both allocation schemes. The absolute values obtained are not of primary interest. Rather the relative impact of the two alternative plans is what is analyzed.

---

[2]The adjustments may be stated as:

$$R_{ij} = 1 - \frac{D_c - D_c'}{D_w} \quad \frac{E_{ij}^{ss}}{E_{ij}^{h}} \quad Q_{ij}$$

where:

R_{ij} = the initial percentage of output of sector i located in region j which remains intact after the curtailment plans go into effect;

D_c = Average number of days curtailed service to interruptible customers under the new allocations plans;

D_c' = Average number of days that customers were on curtailed service in 1973;

D_w = Average total number of working days per year.

E_{ij}^{ss} = Maryland Department of Employment and Social Services employment by firm(s) i located in region j;

E_{ij}^{h} = Regional model employment in sector i located in region j;

Q_{ij} = Output of sector i located in region j.

[3]$(D_c - D'_c)$ was set equal to (49-12) = 37 days.

The results of the pro rata curtailment run of the model are presented in Table 4. The benchmark (no shortage) forecasts of the model are presented in the first row for each of the variables listed. In the second row are the forecasts generated after the imposition of a pro rata curtailment of natural gas. The unconstrained forecast of employment in the state for 1974 is 1.637 million. The pro rata curtailment of natural gas supplies resulted in the employment forecast being reduced to 1.601 million or a reduction in employment of about 2.2 percent. The effect of this one-time shock to the model is diminished in the forecasts for subsequent years.[4] As would be expected, firms will, over time, convert to alternative energy sources. The initial impact of an additional 36,000 unemployed in the State seems high, however.[5]

A simulation with the regional model was also prepared for the FPX priority end-use plan. The results from this simulation are presented in Table 5. The basic adjustments are made to eliminate the production of those industries on interruptible natural gas service, since the proposed FPC regulations require that lower priority users be discontinued before those in a higher category are curtailed. The contrast between the effects of the two plans is dramatic. While the pro rata plan implies a 1.8 percent decrease in output during the first year, the end-use plan results in an 11.0 percent decrease in output. The decrease in output in subsequent years under the end-use plan is not ameliorated as is the case under the pro rata plan and remain at 8.1 percent and 7.7 percent decreases for the years 1975 and 1976, respectively. Increases in the unemployment figures for the end-use plan follow similar trends. Naturally, in the long-run it would be expected that many of the firms which lost their natural gas service under the end-use plan would be able to convert to alternative fuels. However, the curtailment might cause some firms to relocate in other areas of the country which were not experiencing the same degree of fuel shortage.

These figures should be interpreted merely as preliminary results. This regional model was not originally designed to handle the effects of input shortages;

---

[4]This may result because construction and equipment investment exhibit a one-year lag with the highest impact on them occurring in 1975.

[5]The high initial impact on the first year of the forecast and the diminishing impacts on subsequent years may be partly due to the ways that the curtailment plans were incorporated in the model. More reasonable and reliable results might be obtained if the adjustments had been applied to the two years (1973 and 1974), which serve as the base years for the forecast period. Such an adjustment procedure would be more in accord with the empirical basis of the model.

and therefore, more substantive changes in the structure of the model may be required in order to incorporate explicitly the space heating, process and feedstock fuel requirements of each sector. Several points should be briefly discussed concerning assumptions in the model which also merit review.

One issue relates to some simplifying assumptions used in the model, primarily due to the data exigencies. First is the definition of the variable used to represent the value of land. This variable is restricted to agricultural land with no estimate of the value of non-agricultural land being included. The value of the land variable should be respecified. Another consideration which should be evaluated is the possible inclusion of backward linkages between changes in output and its explanatory variables. Also, the lag structure in the model should be estimated and thereby might be improved. Since the model requires such large amounts of data these changes could only be implemented with substantial cost in both time and effort. The reliability of the forecasts could improve substantially with these extensions to the model structure and estimation.

Finally, reductions in output effect investment expenditure patterns. Output, changes in output, and investment represent the main determinants of employment and thereby impose an immediate three-fold pressure on changes in employment. This is one reason why, under the present runs, the impact on employment in the first year of he forecast is possibly high. This result may be attributed in part to the way in which the initial impacts of the natural gas curtailments were introduced into the model. As was explained earlier, the output of each industry was constrained using the employment ratios of the firms affected. This procedure introduces an undesirable rigidity into the model and ignores the instance where firms have different capital intensities than the particular sector where they reside.

VI CONCLUSION

The Maryland Multiregional, Multi-Industry Forecasting Model is used to estimate the impact of the two proposed curtailment plans of natural gas on the economy of the State of Maryland. Its performance seems quite satisfactory. In the absence of good, i.e. long, historical data series and the concomitant ability to run simulations and empirical tests, the evaluation of a comprehensive regional model is not easy. However, the regional model is flexible and yields itself to uses for which it was not originally designed; e.g. it has been applied to an analysis of transportation networks plants. The results presented here indicate that the model can be useful for analyzing input shortages, also. In this context it might be appropriate to aggregate some of the regions and industry sectors, while disaggregating the energy sectors, particularly, coal mining, petroleum mining, petroleum refining, electric utility, gas utility, and transportation. Compared with other extant regional models [3,4,14,15], in terms of the overall methods employed and results achieved, the Harris model deserves high marks and warrants extension and further development.

REFERENCES

Almon, Clopper, M.B. Buckler, L.M. Horwitz, T.C. Reimboldt, 1974, 1985 Inter-industry Forecasts of the American Economy. Lexington, MA: D.C. Heath and Co.

Balestra, P., and M. Nerlove, 1966, "Pooling Cross Section and Time Series Data in the Estimation of a Dynamic Model: The Demand for Natural Gas," Econometrica, Vol. 34.

Bell, Frederick W., 1967, "An Econometric Forecasting Model for a Region," Journal of Regional Science, Vol. 7, No. 2.

Brown, H. James, 1969, "Shift and Share Projections of Regional Economic Growth: An Empirical Test," Journal of Regional Science, Vol. 9.

Conway, Richard S., Jr., 1975, "A Note on the Stability of Regional Interindustry Models," Journal of Regional Science, Vol. 15, No. 1.

Cumberland, John H., W.A. Donnelly, C.S. Gibson, Jr. and C.E. Olson, 1975, "Forecasting Alternative Regional Electric Energy Requirements and Environmental Impacts for Maryland, 1970-1990." Paper presented at the International Regional Science Conference on Energy and Environment, Leuven, Belgium.

Fox, K. and T.K. Kumar, 1965, "The Functional Economic Area," Regional Science Association Papers, Vol. 15.

Green, George R., 1972, "Forecasting State Variables from National Econometric Models," Growth and Change.

Harris, Curtis C., Jr., 1969, "A Multi-Regional, Multi-Industry Forecasting Model." Paper presented at the Regional Science Association Meeting in Santa Monica, California, mimeo.

Harris, Curtis C., Jr., 1974, The Regional Economic Effects of Alternative Highway Systems. Cambridge, MA: Ballinger Publishing Co.

Harris, Curtis C., Jr., 1973, The Urban Economies, 1985: A Multiregional, Multi-Industry Forecasting Model. Lexington, MA.: D.C. Heath and Co.

Harris, Curtis C. and Frank E. Hopkins, 1972, Locational Analysis: An Interregional Econometric Model of Agriculture, Mining, Manufacturing and Services. Lexington, MA: D.C. Heath and Co.

Nissen, David H., D.H. Knapp, W.A. Donnelly, T.E. Eagan and S. Borg, 1975, "The Econometric Regional Demand Model." Federal Energy Administration. Technical Report--FEA EATR 75-18, Preliminary draft.

Olsen, R.J., L.G. Bray, and G.W. Westley, 1974, "The Location of Manufacturing Employment in BEA Economic Areas: A Regression Analysis Using 1950, 1960, and 1970 data." Paper presented at the Southern Regional Association Meetings, Roslyn, VA.

Polenske, Karen R., 1970, A Multiregional Input-Output Model for the United States. Washington, D.C.: U.S. Department of Commerce.

Stilwell, F.J.B. and B.D. Boatwright, "A Method of Estimating Trade Flows," Regional and Urban Studies.

TABLE 3

INPUT DATA FOR PRO RATA CURTAILMENT CASE

Industry Sector		MD County No.	Model Employment	Affected Employment by Sector	Ratio Applied
1	11	1169	25027	62	.99976
2	11	1179	2369	288	.98834
3	14	1169	2994	370	.98815
4	15	1179	107	12	.98925
5	15	1182	330	36	.98954
6	19	1169	942	774	.92121
7	21	1169	2762	749	.97400
8	21	1168	1596	652	.96083
9	24	1168	528	243	.95587
10	32	1169	923	209	.97829
11	33	1167	290	97	.96793
12	35	1169	5064	2599	.95079
13	35	1178	887	94	.98984
14	37	1168	1226	856	.93305
15	37	1169	2827	1187	.95974
16	43	1169	2779	1933	.93330
17	44	1169	3986	2778	.93317
18	45	1169	3461	3461	.90411
19	45	1168	30428	18833	.94065
20	46	1167	954	678	.93185
21	47	1168	489	345	.93235
22	48	1169	8890	318	.99657
23	49	1169	2299	363	.98486
24	50	1169	2965	653	.97888
25	51	1169	1689	211	.98802
26	57	1168	1408	1018	.93067
27	57	1172	3909	2828	.93063
28	58	1169	1388	789	.94549
29	59	1168	323	320	.90500
30	59	1169	168	166	.90525
31	59	1172	54	53	.90589
32	59	1167	178	176	.90519
33	63	1168	670	670	.90411
34	63	1169	158	158	.90411
35	63	1172	586	586	.90411
36	63	1167	613	613	.90411
37	66	1168	3679	3679	.90411
38	66	1169	234	124	.94919
39	66	1167	3186	1688	.94920
40	71	1168	5164	5164	.90411
41	71	1169	4777	4777	.90411
42	71	1167	423	195	.95580
43	72	1168	643	437	.93483
44	72	1169	111	75	.93521
45	74	1172	457	218	.95426

Table 3 (continued)

County Code	County Name
1167	Anne Arundel
1168	Baltimore County
1169	Baltimore City
1172	Carroll
1178	Harford
1179	Howard
1182	Prince Georges

Industry Sectors

11	New Construction
14	Meat Packing
15	Dairy Products
19	Sugar
21	Beverages
24	Fabrics and Yarn
32	Paper Products, exec. Containers
33	Paper Containers
35	Basic Chemicals
37	Drugs, Cleaning and Toilet Items
43	Glass and Glass Products
44	Stone and Clay Products
45	Iron and Steel
46	Copper
47	Aluminum
48	Other Nonferrous Metals
49	Metal Containers
50	Heating, Plumbing, Structural Metals
51	Stampings, Screw Machine Products
57	Metal Working Machinery and Equipment
58	Special Industrial Machinery
59	General Industrial Machinery
63	Electric Apparatus and Motors
66	Communication Equipment
71	Ships, Trains, Trailers, Cycles
72	Instruments and Clocks
74	Miscellaneous Manufactures Products

For a complete list of industry sectors and their SIC components see Harris, [11, pp. 10-12].

TABLE 4

EFFECT OF PRO RATA CURTAILMENT
OF NATURAL GAS SUPPLIES

YEAR	1974	1975	1976
OUTPUT:			
BASE CASE	32704809.	33517302.	36130260.
PRO RATA	32119531.	33194988.	35864956.
PERCENTAGE CHANGE	-1.790	-.962	-.734
EMPLOYMENT:			
BASE CASE	1636940.	1675378.	1713766.
PRO RATA	1600986.	1663063.	1707024.
PERCENTAGE CHANGE	-2.196	-.735	-.393
EARNINGS:			
BASE CASE	15276216.	15662248.	16777522.
PRO RATA	14849038.	15488265.	16657556.
PERCENTAGE CHANGE	-2.796	-1.111	-.715
PERSONAL CONSUMPTION:			
BASE CASE	15779844.	16081082.	17338410.
PRO RATA	15437374.	15907577.	17213853.
PERCENTAGE CHANGE	-2.170	-1.079	-.718
CONSTRUCTION EXPENDITURES:			
BASE CASE	2811490.	2708901.	2826838.
PRO RATA	2811482.	2685398.	2812824.
PERCENTAGE CHANGE	0.0	-.868	-.496
EQUIPMENT EXPENDITURES:			
BASE CASE	1195206.	1261705.	1350757.
PRO RATA	1191837.	1243556.	1328977.
PERCENTAGE CHANGE	-.282	-1.438	-1.612

TABLE 5

EFFECT OF PRO RATA CURTAILMENT
OF NATURAL GAS SUPPLIES

YEAR	1974	1975	1976
OUTPUT:			
BASE CASE	32704809.	33517302.	36130260.
END-USE	29122529.	30813698.	33349995.
PERCENTAGE CHANGE	-18.953	-8.066	-7.695
EMPLOYMENT:			
BASE CASE	1636940.	1675378.	1713766.
END-USE	1453556.	1571331.	1610237.
PERCENTAGE CHANGE	-11.203	-6.210	-6.041
EARNINGS:			
BASE CASE	15276216.	15662248.	16777522.
END-USE	132870083.	14526106.	15622387.
PERCENTAGE CHANGE	-13.021	-7.254	-6.885
PERSONAL CONSUMPTION:			
BASE CASE	15779844.	16081082.	17338410.
END-USE	14081667.	14985444.	16208798.
PERCENTAGE CHANGE	-10.762	-6.813	-6.515
CONSTRUCTION EXPENDITURES:			
BASE CASE	2811490.	2708901.	2826838.
END-USE	2743239.	2594970.	2735118.
PERCENTAGE CHANGE	-2.428	-4.206	-3.245
EQUIPMENT EXPENDITURES:			
BASE CASE	1195206.	1261705.	1350757.
END-USE	1132092.	1082990.	1157518.
PERCENTAGE CHANGE	-5.201	-14.165	-14.306

TECHNOLOGICAL ABATEMENT VS. LOCATIONAL ADJUSTMENT: A TIME-SPACE DILEMMA

Manoucher Parvin and Gus W. Grammas[1]

I INTRODUCTION

The account of material waste as the by-product of production and consumption activities, within the context of input-output or mathematical programming models, was initiated by Leontief [15], Kohn [12, 13], Muller [16], Kneese and Bower [10], and extended by other authors with the most recent additions of Cohen and Hurter [3], Chatterji [2], Parvin [17], and Parvin and Grammas [18]. Discussing the strategies of air pollution control, Kohn [14] suggests the advantages of "locational adjustment" of emitters (or alternatively the relocation of receptors) in contrast to "expensive technological abatement." It is not difficult to point to specific situations where locational adjustment does in fact comprise optimum short-run regional pollution control policy. However, with the existing technological structure and expanding economic activities, the waste assimilative capacity of the ecology for certain by-products has been surpassed. Furthermore, belated public recognition of fossil fuel shortages on the one hand and environmental repercussions of unlimited growth on the other, has introduced new difficulties that locational adjustment cannot help if increased fuel use per unit output is required. The optimization condition achieved by locational adjustment is generally in divergence with temporal and global optimization since the spatial distribution of economic activities is not time invariant. In fact, the danger of widespread use of such a solution is that it diverts scientific and economic resources from the more fundamental and universal solution which is technological abatement.

The same composition and level of pollution can be produced with different mixtures of abatement technology and fuel resources. For example, the use of coal plus abatement facilities may result in similar pollution output as the use of high grade oil and no abatement. In this paper we shall focus our attention on the trade-off possibilities existing between technology, fuel resources and locational adjustment. In particular, the development of a regional model of stationary emitters and the derivation of equilibrium conditions (optimum trade-

[1]The authors would like to express their appreciation to Mr. Carlos Picado-Horta of Columbia University for his assistance in gathering the date and making the extensive computations, to Mr. William Tang of the Department of Air Resrouces of the City of New York for providing much of the basic pollution data and to Ms. Nancy Najarian for her expert assistance in the final preparation of this paper.

offs) of locational adjustment vs. technological abatement and fuel diversification is attempted. Our second objective is to examine the theoretical results in light of practical difficulties and provide data emerging from New York City air pollution control experience.

Pollution control can be achieved by decreasing economic activities, by controlling population growth or by changing the composition of final demand through government punishment (taxation) or reward (subsidy) policies. We assume in our model, however, that such factors or conditions are not interfered with. These assumptions will allow us to focus our attention on technological abatement vs. locational adjustment.

Technological abatement implies one or both of the following: (i) development and use of new factors of production (e.g., higher grade of fuel in power generation) and (ii) change in the process of production (e.g., installation of filters to reduce particle matter or other emissions). One or both of such technological changes will result in a different level and composition of pollution per unit of output of product, for example, kilowatt hours of electricity.

Locational adjustment refers to one of the following possibilities: (i) relocation of a plant with essentially no change in technology, (ii) relocation with some modification in technology and (iii) complete abandonment of a plant and construction of a new plant with new technology. In the pure locational adjustment (i), only the spatial distribution of pollution is altered for the purpose of decreasing the damage cost with the same composition and level of pollution emission.

The distances in this model are a general representation of economic space, such as for example, the economic frictional distance.

The cost of labor and capital input enter the model implicitly through the cost of attaining and maintaining a location specific production technology.

II REGIONAL MODEL FOR A SINGLE INDUSTRY

The following model of the production and dissemination of pollution by a single industry (e.g., electric power) constructs a total cost function. It incorporates the damage to receptors, the cost of technological abatement and the cost of locational adjustment. Equilibrium conditions are then derived with respect to technological abatement and locational adjustment.

Assume that there are I different input fuels ($i = 1, 2, \ldots, I$) used by the industry, that there are J_i different plants using fuel i and that a particular plant uses only one fuel. For purposes of this study, a plant using two or more different input fuels can be designated as two or more different plants. The total number of plants is $\Sigma_{i=1}^{I} J_i$. There are also K different receptor locations, each of which receives pollution from all plants in the industry, and there are N different pollutants produced by the industry. Let

F_{ij} = the amount of fuel i used by plant j

a_{ij} = the amount of fuel i used per unit output of plant j

and

E_{ij} = the total output of plant (i, j).

We have that

$$F_{ij} = a_{ij} E_{ij}$$

and that the total output of the industry is

$$E = \Sigma_{i=1}^{I} \Sigma_{j=1}^{Ji} F_{ij}/a_{ij}. \qquad (1)$$

Also let

b_{ijn} = the amount of pollutant n produced per unit output of plant (i,j)

d_{ijk} = the "distance" between plant (i, j) and receptor location k; this distance can be the usual Euclidean distance between two points or some more general representation of space.

and

$\alpha_n(\Phi, d_{ijk})$ = pollution attenuation function; the fraction of pollutant n produced at (i, j) and received by location u. Φ represents the diffusion characteristics of pollutant n within an environmental regime (Φ itself may be a function of several variables). Stack height, environmental regime condition and, finally, time or flow distribution of pollutant can be incorporated in α_n through Φ or additional parameters for each case. Note that α_n is left unspecified and is to be determined for each particular case. In practice, α_n may be difficult to estimate. Also note that

$$0 \leq \alpha_n \leq 1 \text{ and } \frac{\partial \alpha_n}{\partial d_{ijk}} < 0.$$

Now let

g_{ijn} = the total amount of pollutant n produced by plant (i, j)

$= b_{ijn} \cdot E_{ij}$

and

g_{ijkn} = the total amount of pollutant n received by location k from plant (i, j).

Note that

$$g_{ijkn} = b_{ijn} \cdot E_{ij} \cdot \alpha_n(\Phi, d_{ijk})$$

and that, if

G_{kn} = the total amount of pollutant n received by location k, then

$$G_{kn} = \Sigma_{i=1}^{I} \Sigma_{j=1}^{J_i} b_{ijn} E_{ij} \alpha_n (\Phi, d_{ijk}). \tag{2}$$

Now let

$D_{kn}(G_{kn})$ - average damage cost function (of G_{kn}) per unit of pollutant n per unit time per person in location k. The simplest form that D_{kn} can take is as a linear function of G_{kn}. In practice, however, more complex forms may be required and

π_k = population at location k.

We now define

TD = the total damage cost of pollution by the industry per unit time which, in terms of the input parameters and functions, is

$$TD = \Sigma_{k=1}^{K} \Sigma_{n=1}^{N} \pi_k \cdot G_{kn} \cdot D_{kn} (G_{kn}) \tag{3}$$

In its simplest state (with D_{kn} a linear function), TD is a quadratic form in terms of the input parameter b_{kjn} and the function α_n.

The damage cost of pollution, TD, is decreased by either decreasing b_{ijn} (technological abatement) or increasing d_{ijk} (locational adjustment) or both. Now let

$f_{ijn}(b_{ijn})$ = the total cost associated with attaining and maintaining b_{ijn} at a given magnitude (e.g., the fixed cost of installing a new filter plus the variable cost of increased fuel requirements for the same level of output) and

$h_{ijk}(d_{ijk})$ = the total cost associated with attaining and maintaining a given distance d_{ijk} (e.g., the fixed cost of original relocation plus the variable cost of increased transportation requirements). We assume that

$$\frac{\partial f_{ijn}}{\partial b_{ijn}} < 0 , \quad \frac{\partial^2 f_{ijn}}{\partial b_{ijn}^2} > 0$$

and

$$\frac{\partial h_{ijk}}{\partial d_{ijk}} > 0 , \quad \frac{\partial^2 h_{ijk}}{\partial d_{ijk}^2} > 0 \tag{4}$$

We now define

TC = the total "cost" of pollution production, which, in terms of its three components, is

$$TC = \Sigma_{k=1}^{K} \Sigma_{n=1}^{N} \pi_n G_{kn} D_{kn} (G_{kn}) + \Sigma_{i=1}^{I} \Sigma_{j=1}^{J_i} \Sigma_{n=1}^{N} f_{ijn}(b_{ijn}) + \Sigma_{i=1}^{I} \Sigma_{j=1}^{J_i} \Sigma_{k=1}^{K} h_{ijk}(d_{ijk}) \tag{5}$$

Equilibrium conditions can now be derived with respect to technological abatement and locational adjustment. In order to minimize TC, we have

$$\frac{\partial TC}{\partial b_{ijn}} = \frac{\partial TD}{\partial b_{ijn}} + \frac{\partial f_{ijn}}{\partial b_{ijn}} = 0.$$

$$\frac{\partial TC}{\partial d_{ijk}} = \frac{\partial TD}{\partial d_{ijk}} + \frac{\partial h_{ijk}}{\partial d_{ijk}} = 0$$

and, finally, the equilibrium conditions are

$$\frac{\partial TD}{\partial b_{ijn}} = - \frac{\partial f_{ijn}}{\partial b_{ijn}} \qquad (6)$$

$$\frac{\partial TD}{\partial d_{ijk}} = - \frac{\partial h_{ijk}}{\partial d_{ijk}}$$

for all i, j, k, and n.

In the following two cases, we consider two specific formulations for $D_{kn}(G_{kn})$ and derive the corresponding equilibrium conditions:

Case 1:

Let $D_{kn}(G_{in}) = s_{kn}$, a constant whenever $G_{kn} > 0$. This formulation is primarily presented for purposes of comparison with the next case. However, this formulation may not be totally unrealistic for a small range of G_{kn} or for small values of G_{kn}. For this case, the total damage cost of pollution, TD, is a linear function of G_{kn} and the equilibrium conditions reduce to:

$$\frac{\partial f_{ijn}}{\partial b_{ijn}} = - E_{ij} \cdot \Sigma_{k=1}^{K} \pi_k \cdot s_{kn} \cdot \alpha_n (\Phi, d_{ijk})$$

and

$$\frac{\partial h_{ijk}}{\partial d_{ijk}} = - E_{ij} \cdot \pi_k \cdot \Sigma_{n=1}^{N} \cdot b_{ijn} \cdot s_{kn} \cdot \frac{\partial \alpha_n}{\partial d_{ijk}}$$

Case 2:

Let $D_{kn}(G_{kn}) = s_{kn} + t_{kn}G_{kn}$, a linear function whenever $G_{kn} > 0$. This formulation should be useful for a reasonable range of G_{kn}. For this case, TD is a quadratic function of G_{kn} and the equilibrium conditions reduce to:

$$\frac{\partial f_{ijn}}{\partial b_{ijn}} = - E_{ij} \Sigma_{k=1}^{K} \pi_k [s_{kn} + 2t_{kn}G_{kn}] \alpha_n (\Phi, d_{ijk})$$

and

$$\frac{\partial h_{ijk}}{\partial d_{ijk}} = - E_{ij} \pi_k \Sigma_{n=1}^{N} b_{ijn} [s_{kn} + 2t_{kn}G_{kn}] \cdot \frac{\partial \alpha_n}{\partial d_{ijk}} . \qquad (7)$$

III TECHNOLOGICAL INDICES

Technological progress is achieved by decreasing b_{ijn}, i.e. reducing the amount of pollution, or by decreasing a_{ij}, i.e. increasing productivity, or by some combination of both. The following two indexes attempt to quantify such progress. Starting from the basic equations

$$F_{ij} = a_{ij} E_{ij}$$

and

$$g_{ijn} = b_{ijn} E_{ij}$$

we define

$$R_{ijn} \equiv \frac{b_{ijn}}{a_{ij}} = \frac{g_{ijn}}{F_{ij}}$$

and

$$Tijn \equiv \frac{1}{\sqrt{a_{ij} \cdot b_{ijn}}} = \frac{E_{ij}}{\sqrt{F_{ij} \cdot g_{ijn}}} .$$

R_{ijn} is a pollution emission index; for a given level of a_{ij}, a decrease in R_{ijn} implies a decrease in b_{ijn}. Note that this index is independent of the total output of plant (i,j) and reflects only the efficacy of the pollution abatement procedures instituted by the plant.

T_{ijn}, on the other hand, is a compound technological index; an increase in T_{ijn} implies a decrease in b_{ijn} or a_{ij} or both. Note that this index is also the ratio of useful output to the geometric average of the two inputs (i.e., considering F_{ij} and g_{ijn} as "cost" factors). We expect that further theoretical and empirical exploration of this index, from both the technological and economic point of view, will result in even more revealing considerations.

IV TECHNOLOGICAL ABATEMENT VS. LOCATIONAL ADJUSTMENT: DISCUSSION

The above model gives the theoretical trade-off equilibrium condition between the two approaches. However, the simplifying assumptions made, by necessity, abstract away some critical parts of aspects of the problem which can be broadly classified as "externalities." The cost functions stipulated in the abova analysis could not possibly incorporate all such externalities fully. Two types of externalities, short-run and long-run, are associated with the two methods of pollution control.

In the short-run, locational adjustment implies spatial redistribution of waste "pollution" disposal. This leads to a redistribution of economic "bads" for which the newly affected section of the population must be, and theoretically can be, compensated for although it is difficult to develop the criteria that would satisfy all newly affected consumers. Given the increased consciousness of the populace concerning environmental pollution, locational adjustment may take a long time due to resistance (political and legal) of the host community. Such problems can be avoided if "filtering" devices are installed in present locations.

Locational adjustment lacks flexibility in conjunction with composition of pollution. Recent studies show that synergistic pollutants, such as copper and lead compounds, in combination cause a greater damage cost than the sum of components independently, while antagonistic pollutants, such as calcium and lead compounds, tend to neutralize each other (e.g., Crossan [4], Dorn, et. al. [5] and Schneider [21]). Methods of technological abatement can take advantage of such discoveries to diminish damage cost by altering the pollution composition as well as the level of pollution.

Each locational adjustment contains more or less similar problems, procedures, and obstacles and thus is not subject to mass production economies or the like. However, technological abatement devices or methods, once developed, can be used in many locations with perhaps diminishing cost of manufacturing, installation and maintenance due to increased productivity resulting from learning and experience.

In the long-run, locational adjustment does not affect the global environmental problem which is, specie-wise, a more important problem. Given the irreversibility of many production and consumption activities and limited assimilative capacity of the planet, locational adjustment appears as a transitory and elusive solution. Only in case of new plants, given the available abatement technology, should locational factors be given serious consideration. Obviously, optimization criteria of new installations are less subject to past history than locational adjustment of operating plants. Finally, population growth and the accompanying spatial spreading may in a short span of time render an optimum location solution less attractive.

It is due to above and other difficulties that Consolidated Edison of New York has chosen technological abatement rather than locational adjustment. In fact, experience of the past several years indicates great legal-political difficulties for even the installation of new plants in more locationally appropriate places.

V THE AIR POLLUTION CONTROL EXPERIENCE OF CONSOLIDATED EDISON

Within the last decade, much attention has been given to the environmental impact of electrical power production. Air pollution, the major form of pollution resulting from electric power generation, is discussed by Wolozin [26], Stern [22], and Kneese, et. al. [11]. Fabricant and Hallman [8], Garvey [9], and Esposito [7] provide thorough discussions of the effects of the pollutants produced on human health and the natural environment. Procedures for calculating some of the costs of air pollution have been developed by Ridker [20] and Watson [25].

The pollution attenuation function, α_n, represents the amount of pollutant n that actually reaches a population center. Turner [23, 24] discusses the use of diffusion models in air pollution and his work, among others, has led to the construction of the Federal Environmental Protection Agency's Air Quality Display Model [1]. In this model for particulate matter emissions, not only are emission rates considered, but the position of the source with respect to the prevailing winds and the height of the source above the ground are considered as well. However, the exact specification of α_n, for different pollutants, still remains a difficult problem.

The electric industry is of great importance to the economy due to its interdependence with commercial, industrial, residential sectors and ecology concerns. The externalities resulting from electric power generation can cause severe illness and disease, contaminate valuable crops, destroy property, and create negative psychological and aesthetic effects.

The damage cost, D_{kn}, is also difficult to specify exactly. Eisenbud [6] states that the average economic loss due to air pollution in large urban areas is $65 per person per year, but that there have been no studies which apportion these costs among the various sources of air pollution. Rabow [19] outlines a systematic approach for determining the costs of loss in health or well-being due to air pollution and illustrates this approach by describing a scheme for determining the cost of excess mortality due to air pollution. In general, D_{kn} must take into account the loss of health in man and animals and the damage to plants and property.

Coal, oil and gas are the input fuels utilized by Consolidated Edison for electric power generation in New York City. (Note that no nuclear power plants are located within city limits.) A severe decline in coal usage has occurred due to its undesirable sulfur content; on the other hand, oil and gas consumption has soared (Table 1). The low polluting properties of oil and gas (especially with respect to sulfur dioxide and particulates) are responsible for the rapid growth in demand for these fuels. Lengthy construction periods and red tape regarding site and installation approval have precluded the use of nuclear fuel in New York City.

In New York City, the five major air pollutants produced are sulfur dioxide, particulate matter, nitrogen oxides, hydrocarbons and carbon monoxide. Of these, it appears that the two most dangerous are sulfur dioxide and particulates.

The major sources of pollution production are classified mobile or stationary. Mobile sources, such as cars, planes and trains, are largely responsible for the emission of hydrocarbons and carbon monoxide. Stationary sources, such as fixed industrial (e.g., Consolidated Edison), commercial and residential sites, are most responsible for producing sulfur dioxide and particulates. Both share the burden of nitrogen oxides.

In the production of electricity, negligible amounts of carbon monoxide and hydrocarbons are produced. Thus, our main attention will focus on the major contributions, made by Consolidated Edison, to air pollution levels in New York with respect to sulfur dioxide, particulates and nitrogen oxides.

In late 1970, Eisenbud [6] described the serious problems of air pollution that New York City faced. Since 1965, however, a significant decline in the total production of sulfur dioxide, particulate matter and carbon monoxide has occurred. With specific reference to Consolidated Edison, there has been an 85% decline in sulfur dioxide output, a 74% decline in total particulate matter output, and a 21% decline in total carbon monoxide produced (Table 2). Nitrogen oxides still remain a problem.

In order to show how each Consolidated Edison plant in New York City has contributed to the decline in pollution production, the coefficients F_{ij} (Tables 3-5) and g_{ijn} (Tables 6-9) are presented for the years 1965, 1972 and 1973. In addition, the pollution emission index, R_{ijn} (Table 10), for fuel oil #6 is presented for these years (note that this index has been decreasing over time for the major pollutants). At present, a determination of all the a_{ij} and b_{ijn} is not possible since the required data are lacking. Such a determination is left for future study.

The gains made in reducing total sulfur dioxide and particulate matter emissions are in evidence throughout the city. A decline in sulfur dioxide levels has occurred in each borough. For the city as a whole, federally decreed pollutant levels have been attained for sulfur dioxide (Table 11) and are being approached for particulate matter (Table 12).

Consolidated Edison has achieved its greatly reduced output of pollutants by concentrating on technological abatement. Utilization of low sulfur fuels is the single most important factor responsible for the decline in sulfur dioxide output. The installation of mechanical and electrostatic precipitators has profoundly relieved the particulate matter problem (these highly efficient devices prevent particulates from leaving the stacks and entering the atmosphere). Use of stack liners and increased efficiency in fuel combustion have helped control nitrogen oxides and reduce hydrocarbon emissions. Finally, the updating of older facilities and the utilization of new and technologically more efficient plants have also aided in the reduction of hydrocarbons.

VI SUMMARY AND CONCLUSIONS

A simple mathematical model has been presented which yields the theoretical equilibrium conditions with respect to technological abatement and locational adjustment. The model is a static one and, thus, does not yield the dynamic optimization conditions insofar as global environmental pollution is concerned.

It is argued that locational adjustment is, in a long-run and dynamic sense, inferior to technological abatement. In this regard, data are presented which are pertinent to the experience with pollution control of Consolidated Edison of New York. It is seen that, due to a variety of obstacles, the option of locational adjustment has not been exercised. Indeed, sites for new plants, incorporating even the most advanced pollution abatement technology, are extremely difficult to obtain.

Further investigation in this area is required before a complete resolution of the issues is obtained. At present, sufficient data on the benefits and costs of locational adjustment do not exist. It is our contention, however, that in the final analysis locational adjustment should be considered only as an exceptional, transitory and regional solution to pollution control problems.

Insofar as welfare economics and distribution of pollution responsibilities are concerned the following important points must be emphasized. The average cost of a pollutant is an increasing function of the level of that pollutant for a given state of technology and spatial distribution of emitters and receptors. Thus the larger consumers of electricity inflict a higher external cost not simply because of scale but because of increasing costs of pollution. Accordingly, the price charged to large users must be higher to compensate for this social external cost. It is interesting to note that the reverse condition prevails today in New York City. A lower unit price is charged to large users of electrical power thus providing no incentive for reduced use of electrical energy.

REFERENCES

Air Quality Display Model, 1972. National Technical Information Service, PB-189-194.

Chatterji, M., 1975. "A Dynamic Balanced Regional Input-Output Model of Pollution Control," Environment and Planning A, Vol. 7, No. 1, pp. 21-34.

Cohen, A.S. and A.P. Hurter, 1974. "An Input-Output Analysis of the Costs of Air Pollution Control," Management Science, Vol. 21, No. 4, pp. 453-461.

Crossan, A.B., 1973. "The Raritan River 1972: Study of the Effect of the American Cyanamid Company on the River Eco System." Unpublished Ph.D. dissertation, Graduate Program in Geophysical Fluid Dynamics, Rutgers University, New Brunswick, N.J.

Dorn, R.S., J. Pierce II, G.H. Chase, and P. Phillips, 1972. "Study of Lead, Copper, Zinc, Cadmium Contamination of Food Chains of Man," National Technical Information Service, PB-223-018.

Eisenbud, M., 1970. "Environmental Protection in the City of New York," Science, Vol. 170, pp. 706-712.

Esposito, J.C., 1970. Vanishing Air: The Ralph Nadar Study Group Report on Air Pollution. New York: Grossman Publishers.

Fabricant, N. and R.N. Hallman, 1971. Toward a Rational Power Policy: Energy, Politics and Pollution. New York: George Braziller, Inc.

Garvey, G., 1972. Energy, Ecology, Economy. New York: W.W. Norton and Company, Inc.

Kneese, A.V. and B.T. Bower, 1968. Managing Water Quality: Economics, Technology, Institutions. Baltimore: Johns Hopkins Press.

Kneese, A.V., R. Ayres and R. d'Arge, 1970. Economics and the Environment: A Materials Balance Approach. Washington: Resources of the Future, Inc.

Kohn, R.E., 1971. "Application of Linear Programming to a Controversy on Air Pollution Control," Management Science, Vol. 17, No. 10 (1971), pp. B609-B621.

Kohn, R.E., 1971. "Optimal Air Quality Standards," Econometrica, Vol. 39, pp. 983-995.

Kohn, R.E., 1974. "Industrial Location and Air Pollution Abatement," Journal of Regional Science, Vol. 14, No. 1, pp. 55-63.

Leontief, W., 1970. "Environmental Repercussions and Economic Structures: An Input-Output Approach," Review of Economics and Statistics, Vol. 52, No. 3, pp. 262-271.

Muller, F., 1973. "An Operational Mathematical Programming Model for the Planning of Economic Activities in Relation to the Environment," Socio-Economic Planning Sciences, Vol. 7, pp. 123-138.

Parvin, M., 1974. "Pollution Control Policies and Strategies: A Mathematical Programming Approach," presented at the European Meeting of the Regional Science Association, Karlsruhe, Germany.

Parvin, M. and G.W. Grammas, 1975. "Optimization Models for Environmental Pollution Control: A Synthesis," presented at the ORSA/TIMS National Meeting, Chicago, April.

Rabow, G., 1973. "Cost of Pollution," IEEE Transactions on Systems, Man, and Cybernetics, Vol. SMC-3, No. 3, pp. 275-276.

Ridker, R.G., 1967. Economic Costs of Air Pollution: Studies in Measurement. New York: F.A. Praeger Inc.

Schneider, R.F., 1971. "The Impact of Various Heavy Metals on the Aquatic Environment," National Technical Information Service, PB-214-562, February.

Stern, A.C., 1968. Air Pollution, 3 vols. New York: Academic Press.

Turner, D.B., 1964. "A Diffusion Model for an Urban Area," Journal of Applied Meteorology, Vol. 3, No. 1, pp. 83-91.

Turner, D.B., 1969. Workbook of Atmospheric Dispersion Estimates, National Air Pollution Control Administration, Cincinnati.

Watson, W.D., 1972. "Cost of Air Pollution in the Coal-Fired Electric Power Industry," Quarterly Review of Economics and Business.

Wolozin, H. (ed.), 1960. The Economics of Air Pollution. New York: W.W. Norton and Company, Inc.

TABLE 1

ELECTRICITY GENERATED AND FUEL USED BY
CONSOLIDATED EDISON FOR ELECTRICITY GENERATION
IN NEW YORK CITY, 1965 - 1973

Year	Net Electricity Generated (10^6 kwh)	Coal (Tons)	Oil # 6 and #2 (10^3 Gal)	Gas (10^6 CF)
1965	24,985.988	4,939,984	699,128	57,700
1966	26,740.153	4,639,572	861,700	58,694
1967	27,561.258	4,916,836	807,836	70,497
1968	28,192.234	4,461,949	864,523	81,284
1969	29,249.206	3,758,047	1,148,032	79,066
1970	30,665.886	2,583,601	1,595,945	78,578
1971	29,440.096	1,258.923	1,831,384	72,334
1972	27,991.386	145,316	1,918,601	59,583
1973	26,470.443	-	1,836,899	50,104

Source: Annual Reports of Consolidated Edison Company of New York to the State of New York Public Service Commission, 1965-1973.

TABLE 2

ELECTRICITY GENERATED AND AIR POLLUTION EMISSIONS FROM ELECTRICITY GENERATED BY CONSOLIDATED EDISON IN NEW YORK CITY, 1965, 1972, 1973

Year	Net Electricity Generated (10^6 kwh)	Particulate Matter	Sulfur Dioxide	Hydro-carbons	Nitrogen Oxides	Carbon Monoxide
		Tons/Year				
1965	24,985.988	30,776	276,915	838	98,475	4,009
1972	27,991.386	10,377	58,110	244	119,909	3,454
1973	26,470.443	8,084	41,779	208	111,510	3,180

Sources: (1) Annual Reports of Consolidated Edison Company of New York, Inc. to The State of New York Public Service Commission, 1965, 1972 and 1973.

(2) Department of Air Resources, Bureau of Technical Services, The City of New York.

TABLE 3

ELECTRICITY GENERATED AND FUEL USED BY CONSOLIDATED EDISON FOR ELECTRICITY GENERATION IN NEW YORK CITY FOR 1965

Station Designation	Net Electricity Generated (10^6 kwh)	Fuel 1 Coal (Tons)	Fuel 2 Oil #6 (10^3 Gal)	Fuel 3 Oil #2 (10^3 Gal)	Fuel 4 Gas (10^6 CF)
EAST RIVER (Man.)	2,306.673	723,163	83,764	-	124
SHERMAN CREEK (Man.)	645.106	313,015	1,234	-	429
WATERSIDE (Man.)	2,224.633	455,840	1,350	-	17,231
74th STREET (Man.)	619.825	295,708	2,103	-	-
59th STREET (Man.)	491.346	133,757	16,134	-	14
HELL GATE (Bx.)	1,506.092	-	122,436	-	9,546
HUDSON AVE. (Bk.)	2,393.556	297,634	192,524	-	-
KENT AVE. (Bk.)	280,011	134,646	-	-	792
RAVENSWOOD (Q.)	5,930.746	-	275,625	-	16,714
ASTORIA (Q.)	6,612.500	1,916,568	1	-	12,850
ARTHUR KILL (S.I.)	1,975.500	669,653	3,957	-	-

Source: Annual Report of Consolidated Edison Company of New York, Inc. to the State of New York Public Service Commission, 1965.

TABLE 4

ELECTRICITY GENERATED AND FUEL USED BY CONSOLIDATED EDISON FOR ELECTRICITY GENERATION IN NEW YORK CITY FOR 1972

Station Designation	Net Electricity Generated (10^6 kwh)	Fuel 1 Coal (Tons)	Fuel 2 Oil #6 (10^3 Gal)	Fuel 3 Oil #2 (10^3 Gal)	Fuel 4 Gas (10^6 CF)
EAST RIVER (Man.)	2,366.286	-	86,662	-	17,664
WATERSIDE (Man.)	1,454.864	-	106,023	-	9,008
74th STREET (Man.)	445.323	-	46,208	62	-
59th STREET (Man.)	590.230	-	64,261	56	9
HELLGATE (Bx.)	1,388.261	-	144,699	-	3,156
HUDSON AVE. (Bk.)	1,121.519	-	152.199	-	-
RAVENSWOOD (Q.)	9,494.875	-	601,276	-	12,796
ASTORIA (Q.)	7,510,958	5,584	478,766	-	16,950
ARTHUR KILL (S.I.)	3,619.070	139,732	236,369	2,020	-

Source: Annual Report of Consolidated Edison Company of New York, Inc. to the State of New York Public Service Commission, 1972, pp. E-38, E-39, and E-39A.

TABLE 5

ELECTRICITY GENERATED AND FUEL USED BY CONSOLIDATED EDISON FOR ELECTRICITY GENERATION IN NEW YORK CITY FOR 1973

Station Designation	Net Electricity Generated	Fuel 1	Fuel 2	Fuel 3	Fuel 4
		Coal	Oil #6	Oil #2	Gas
	(10^6 kwh)	(Tons)	(10^3 Gal)	(10^3 Gal)	(10^6 CF)
EAST RIVER (Man.)	1,798.712	-	85,993	-	9,883
WATERSIDE (Man.)	1,438.988	-	85,558	-	10,393
74th STREET (Man.)	446.471	-	54,951	33	-
59th STREET (Man.)	602.892	-	67,121	16	17
HELLGATE (Bx.)	215.816	-	37,757	-	38
HUDSON AVE. (Bk.)	1,343.840	-	181,020	-	-
RAVENSWOOD (Q.)	8,821.792	-	560,134	-	11,132
ASTORIA (Q.)	8,264.453	-	509.201	-	18,641
ARTHUR KILL (S.I.)	3,537.479	-	254,017	1,098	-

Source: Annual Report of Consolidated Edison Company of New York, Inc. to the State of New York Public Service Commission, 1973, pp. E-38, E-39, and E-39A.

TABLE 6

AMOUNT OF POLLUTANT n PRODUCED AT EACH PLANT (1,j) FOR FUEL 1 (COAL)

Station Designation	Year	Particulate Matter	Sulfer Dioxide	Hydro-carbons	Nitrogen Oxides	Carbon Monoxide
		(Tons/Year)				
EAST RIVER(Man.)	1965	2,490	23,770	108	6,508	362
	1972	-	-	-	-	-
	1973	-	-	-	-	-
WATERSIDE(Man.)	1965	1,782	21,479	68	4,103	228
	1972	-	-	-	-	-
	1973	-	-	-	-	-
74th ST. (Man.)	1965	1,068	7,079	44	2,661	148
	1972	-	-	-	-	-
	1973	-	-	-	-	-
59th ST.(Man.)	1965	551	3,719	20	1,204	67
	1972	-	-	-	-	-
	1973	-	-	-	-	-
HELLGATE (Bx.)	1965	-	-	-	-	-
	1972	-	-	-	-	-
	1973	-	-	-	-	-
HUDSON AVE.(Bk.)	1965	1,037	9,840	45	2,679	149
	1972	-	-	-	-	-
	1973	-	-	-	-	-
RAVENSWOOD(Q.)	1965	-	-	-	-	-
	1972	-	-	-	-	-
	1973	-	-	-	-	-
ASTORIA (Q.)	1965	6,110	45,154	285	17,249	958
	1972	31	94	0.84	50	2.79
	1973	-	-	-	-	-
ARTHUR KILL(S.I.)	1965	1,992	13,360	100	6,027	335
	1972	736	2,389	21	1,258	70
	1973	-	-	-	-	-

Source: Department of Air Resources, Bureau of Technical Services, The City of New York.

TABLE 7

AMOUNT OF POLLUTANT n PRODUCED AT EACH PLANT (2,j) FOR FUEL 2 (OIL #6)

Station Designation	Year	Particulate Matter	Sulfur Dioxide	Hydro-carbons	Nitrogen Oxides	Carbon Monoxide
		(Tons/Year)				
	1965	1,633	15,770	8.38	4,398	126
EAST RIVER(Man.)	1972	498	3,293	8.67	4,550	130
	1973	361	1,884	8.60	4,515	129
	1965	27	261	0.14	71	2.03
WATERSIDE (Man.)	1972	631	4,271	10.60	5,566	159
	1973	359	1,855	8.56	4,492	128
	1965	42	405	0.21	110	3.15
74th STREET(Man.)	1972	217	1,217	4.62	2,426	69
	1973	243	1,278	5.50	2,885	82
	1965	315	3,024	1.61	847	24
59th STREET(Man.)	1972	302	1,692	6.43	3,374	96
	1973	305	1,665	6.71	3,524	101
	1965	2,418	23,150	12.24	6,428	184
HELLGATE (Bx.)	1972	720	4,258	14.47	7,597	217
	1973	167	878	3.78	1,982	57
	1965	3,754	36,089	19.25	10,108	289
HUDSON AVE.(Bk.)	1972	852	5,545	15.22	7,990	228
	1973	760	3,924	18.10	9,504	272
	1965	5,513	53,113	27.56	14,470	413
RAVENSWOOD (Q.)	1972	2,856	16,299	60.13	31,567	902
	1973	2,353	12,142	56.01	29,407	840
	1965	0.023	0.20	0.0001	0.053	0.0002
ASTORIA (Q.)	1972	2,226	12,241	47.88	25,135	718
	1973	2,286	12,237	50.92	26,773	764
	1965	77	735	0.40	208	5.94
ARTHUR KILL(S.I.)	1972	1,158	6,772	23.64	12,409	355
	1973	1,123	5,910	25.40	13,336	381

Source: Department of Air Resources, Bureau of Technical Services, The City of New York.

TABLE 8

AMOUNT OF POLLUTANT n PRODUCED AT EACH PLANT (3,j) FOR FUEL 3 (OIL #2)

Station Designation	Year	Particulate Matter	Sulfur Dioxide	Hydro-carbons	Nitrogen Oxides	Carbon Monoxide
		(Tons/Year)				
	1965	-	-	-	-	-
EAST RIVER(Man.)	1972	-	-	-	-	-
	1973	-	-	-	-	-
	1965	-	-	-	-	-
WATERSIDE(Man.)	1972	-	-	-	-	-
	1973	-	-	-	-	-
	1965	-	-	-	-	-
74th STREET(Man.)	1972	0.02	0.70	0.01	3.26	0.02
	1973	0.01	0.33	0.003	1.73	0.01
	1965	-	-	-	-	-
59th STREET(Man.)	1972	0.02	0.68	0.005	2.94	0.01
	1973	0.01	0.17	0.001	0.84	0.004
	1965	-	-	-	-	-
HELLGATE(Bx.)	1972	-	-	-	-	-
	1973	-	-	-	-	-
	1965	-	-	-	-	-
HUDSON AVE.(Bk.)	1972	-	-	-	-	-
	1973	-	-	-	-	-
	1965	-	-	-	-	-
RAVENSWOOD (Q.)	1972	-	-	-	-	-
	1973	-	-	-	-	-
	1965	-	-	-	-	-
ASTORIA (Q.)	1972	-	-	-	-	-
	1973	-	-	-	-	-
	1965	-	-	-	-	-
ARTHUR KILL(S.I.)	1972	0.71	20	0.17	106	0.51
	1973	0.38	10.13	0.09	58	0.27

Source: Department of Air Resources, Bureau of Technical Services, The City of New York.

TABLE 9

AMOUNT OF POLLUTANT n PRODUCED AT EACH PLANT (4,j) FOR FUEL 4 (GAS)

Station Designation	Year	Particulate Matter	Sulfur Dioxide	Hydro-carbons	Nitrogen Oxides	Carbon Monoxide
		(Tons/Year)				
	1965	0.31	0.04	0.06	37	1
EAST RIVER(Man.)	1972	44	5.30	8.83	5,299	150
	1973	25	2.96	4.94	2,965	84
	1965	43	5.17	8.62	5,169	146
WATERSIDE (Man.)	1972	23	2.70	4.50	2,702	77
	1973	26	3.12	5.20	3,118	88
	1965	-	-	-	-	-
74th STREET(Man.)	1972	-	-	-	-	-
	1973	-	-	-	-	-
	1965	0.04	0.004	0.007	4.2	0.12
59th STREET(Man.)	1972	0.02	0.003	0.005	2.7	0.08
	1973	0.04	0.01	0.01	5.1	0.14
	1965	24	2.86	4.77	2,864	81
HELLGATE (Bx.)	1972	7.89	0.95	1.58	947	27
	1973	0.10	0.01	0.02	11.7	0.42
	1965	-	-	-	-	-
HUDSON AVE.(Bk.)	1972	-	-	-	-	-
	1973	-	-	-	-	-
	1965	42	5	8.36	5,014	142
RAVENSWOOD (Q.)	1972	32	3.84	6.40	3,839	109
	1973	28	3.34	5.57	3,340	95
	1965	32	3.86	6.43	3,855	109
ASTORIA (Q.)	1972	42	5.09	8.48	5,085	144
	1973	47	5.59	9.32	5,592	158
	1965	-	-	-	-	-
ARTHUR KILL(S.I.)	1972	-	-	-	-	-
	1973	-	-	-	-	-

Source: Department of Air Resources, Bureau of Technical Services, The City of New York.

TABLE 10

POLLUTION EMISSION INDEX (R_{2jn})

FUEL 2 (OIL #6)

Station Designation	Year	Particulate Matter	Sulfur Dioxide	Hydro-carbons	Nitrogen Oxides	Carbon Monoxide
		(Tons/10^3 Gal)				
	1965	0.01950	0.18829	0.00010	0.05251	0.00150
EAST RIVER(Man.)	1972	0.00575	0.03800	0.00010	0.05251	0.00150
	1973	0.00420	0.02168	0.00010	0.05232	0.00150
	1965	0.01990	0.19233	0.00010	0.05233	0.00149
WATERSIDE(Man.)	1972	0.00595	0.04029	0.00010	0.05250	0.00150
	1973	0.00420	0.02168	0.00010	0.05250	0.00150
	1965	0.01999	0.19275	0.00010	0.05235	0.00150
74th STREET(Man.)	1972	0.00469	0.02634	0.00010	0.05250	0.00149
	1973	0.00442	0.02326	0.00010	0.05250	0.00149
	1965	0.01952	0.18741	0.00010	0.05250	0.00149
59th STREET(Man.)	1972	0.00470	0.02633	0.00010	0.05251	0.00149
	1973	0.00454	0.02481	0.00010	0.05250	0.00150
	1965	0.01975	0.18909	0.00010	0.05250	0.00149
HELLGATE (Bx.)	1972	0.00498	0.02943	0.00010	0.05250	0.00150
	1973	0.00442	0.02325	0.00010	0.05249	0.00151
	1965	0.01950	0.18746	0.00010	0.05251	0.00150
HUDSON AVE.(Bk.)	1972	0.00560	0.03643	0.00010	0.05250	0.00150
	1973	0.00420	0.02168	0.00010	0.05250	0.00150
	1965	0.02000	0.19272	0.00010	0.05250	0.00150
RAVENSWOOD (Q.)	1972	0.00475	0.02711	0.00010	0.05250	0.00150
	1973	0.00420	0.02168	0.00010	0.05250	0.00150
	1965	-	-	-	-	-
ASTORIA (Q.)	1972	0.00465	0.02557	0.00010	0.05250	0.00150
	1973	0.00449	0.02403	0.00010	0.05258	0.00150
	1965	0.01949	0.18603	0.00010	0.05264	0.00151
ARTHUR KILL(S.I.)	1972	0.00490	0.02865	0.00010	0.05250	0.00150
	1973	0.00442	0.02327	0.00010	0.05250	0.00150

Source: (1) Annual Reports of Consolidated Edison Company of New York, Inc. to The State of New York Public Service Commission, 1965, 1972 and 1973.

(2) Department of Air Resources, Bureau of Technical Services, The City of New York.

TABLE 11

SULFUR DIOXIDE ARITHMETIC AVERAGE (PPM) FOR NEW YORK CITY

	1969	1970	1971	1972	1973	1974
CITY WIDE	.080	.077	.040	.024	.026	.022
BRONX	.079	.077	.041	.026	.026	.019
BROOKLYN	.081	.174	.036	.021	.023	.022
MANHATTAN	.110	.093	.052	.032	.036	.030
QUEENS	.066	.074	.034	.019	.025	.017
RICHMOND (S.I.)	.066	.074	.036	.021	.022	.021

(Air Quality Standard is .03 PPM)

TABLE 12

PARTICULATE MATTER GEOMETRIC AVERAGE ($\mu g/m^3$) FOR NEW YORK CITY

	1969	1970	1971	1972	1973	1974
CITY WIDE	98	104	105	82	82	79
BRONX	101	105	105	91	80	76
BROOKLYN	89	104	110	78	78	77
MANHATTAN	124	124	113	89	91	89
QUEENS	83	97	98	75	74	71
RICHMOND (S.I.)	94	102	97	77	85	81

(Air Quality Standard is 75 $\mu g/m^3$)

Source: Department of Air Resources, Bureau of Technical Services, The City of New York.

COMMENTS ON PARVIN AND GRAMMAS PAPER

S. De Kock

My remarks concerning Mr. Parvin and Mr. Grammas interesting paper are supplementary rather than critical. I agree with their main point that in the long run and from a global point of view technological abatement is in general superior to locational adjustments. But it would be nice if this could be proven more convincingly. The equilibrium conditions, though essential for any evaluation of relocation as against technological abatement, do not immediately reveal this alleged superiority.

Neither are the data emerging from the experience with New York City air pollution control of any help. As long as these results are not compared with the ones that would have emerged from relocating Consolidated Edison, one cannot generate conclusions about the relative superiority of technological abatement in the real world. Furthermore, some of their main arguments against locational adjustment in the second part of the paper are rather tentative, and can in some cases as well be applied in favour of relocation. I will now try to substantiate this point. First, take the case of possible positive effects of particular compositions of different pollutants when technological abatement is imposed. In this connection the authors argue that some pollutants, if emitted within the same environment, neutralize each other, and that a policy of technological abatement can take advantage of such discoveries to diminish damage costs by altering the composition of pollution. They contend however, that in this case locational adjustment lacks flexibility. But I think that relocating different industries in some optimal way with due regard to the composition of pollutants and their respective abatement costs, could as well have this positive neutralization effect. The reason therefore is that it is more likely that two different kinds of pollutants are emitted by different industries rather than by, say, two different plants within one industry.

Secondly, the authors argue that the optimal condition for locational adjustment is generally in divergence with intertemporal and global optimization. In this connection I think that the authors could have made a distinction between two kinds of pollution, i.e. pollutants for which the assimilative capacity of the ecology is limited, and those for which there are no such constraints. It is indeed true that for the first kind of pollutants relocation cannot be a definitive solution, although also this should first be proven. For the second kind of pollutants however, relocation will, if cheaper than technological abatement, be a superior control method. I was thinking about, e.g. an industry with a stenchy output like pig raising. Once pig raising is relocated, all damage costs of pollution disappear in the short run and, I think, as well in the long run, and also from a global point of view, because it seems hardly likely that production will soar in the next decade to such an extent that the whole of Belgium would be "pig-perfumed". (Of course, this would be less likely to the extent that population in Belgium would drastically increase.) But, one of the assumptions of the authors was, that, and I quote, "such factors or conditions are not interfered with".

I think there are other points which could be mentioned in favour of locational adjustment. First, one can make the additional assumption that a high concentration of industries near a populated area increases the costs which one industry imposes on others. Kohn, in a recent article on locational adjustment, argues that the social costs associated with the allowable pollution levels are the abatement costs which each polluter's presence imposes on other polluters. If these kinds of interpendencies are taken into account, it could be that the social costs, which merely result from a high industrial concentration and *a fortiori* from locational factors, could more adequately be removed or diminished with locational adjustment than with technological abatement.

A very interesting feature of the model is that it can be applied to other than air pollutants as well. As a particular interesting application consider the combination of congestion and air pollution caused by cars. In dealing with these simultaneous pollutants, i.e. congestion and carbon monoxide, locational adjustment could in particular situations be advisable. Indeed, relocation would not only solve the air pollution problem but it would also constitute the only alternative way of coping with the congestion problem. What I want to stress here is that for many real world problems locational adjustment will certainly remain a very useful tool. Another remark is again related to the assimilative capacity of the ecology. I think one important control method which is frequently applied in Belgium cannot be overlooked in this kind of models, i.e. laying out green belts or even woods in polluted areas. This measure has the effect of eliminating damage costs related to pollution emissions and could in some cases be cheaper than technological abatement or relocation.

This brings me to a more general point. I believe that it may be misleading to concentrate on two kinds of control methods and assume others away. The reason lies in the fact that there are important interrelationships between the relative effectiveness of different control methods if applied simultaneously. It is, e.g. evident that the effectiveness of locational adjustment is highly dependent on population policy.

This brings me to an even more general remark. The authors stress the importance of intertemporal and global considerations regarding pollution controls. But in doing so they use a static and partial analysis. It would certainly pay to set up a global dynamic model for a particular region built into this model: more than one industry, the possible future effects of the limited assimilative capacity of the ecology and the prospective growth of population and gross regional product (GRP). Perhaps the equilibrium conditions for locational adjustment and technological abatement would turn out to be totally different, and perhaps more operational than the ones obtained in the paper here. Such an attempt would certainly enhance the relevancy of their elegant, but too narrow-scoped model.

Finally I have two minor remarks left. First, I wonder whether the cost curves would turn out to be sufficiently well-behaved in the real world. If this would not be the case, as I presume, the operational significance of the derived trade-off between locational adjustment and technological abatement would be weakened. The second point is related to the data emerging from New York air pollution control experience. The authors argue that more oil and gas are now being used instead of coal because of less damaging effect on the environment, so they claim. But, could it not be that the industries use more oil and gas merely because these forms of energy have become relatively cheaper vis-a vis coal?

FROM WAGES TO BADLY CIRCULATING RENT

A. R. G. Heesterman

I INTRODUCTION

I had some reservations against Meadow's first Rerpot to the Club of Rome[1] but on the whole I agree with the more detailed and qualified second report by Masarovic and Pestel.[2]

The period of economic growth between 1945 and 1973 was characterized by increasing productivity of labour and increased demands on natural and environmental resources. The associated valuation of economic resources imputed the lion's share of the value of the resulting product, to technically trained human labour and to the capital needed to finance the production of new more productive machine. That valuation makes sense, if we assume that natural resources are and will remain abundant. It then matters how long natural and environmental resources will remain undestroyed. I submit that, to the extent that we are concerned about the well-being of future generations, the relatively low valuation of natural resources which prevailed in the sixties, was an under-valuation. The problem of the valuation of natural resources is not new. It is more a question of the subject having laid dormant during the era of technical progress.

II RENTS ON NATURAL RESOURCES

II.1 Rents on indestructible resources

The classical term for the imputed value of a gift of nature is the word rent. The term rent looms large in the older economics textbooks.[3] A rent is a payment of a sum of money made to the owner of some resource, for the use of this resource in production. In the case of indestructible resources, this payment will normally be of a recurrent nature i.e., per year, per month, etc. The payment of land by a farmer to a land-owner is the obvious example.

A rent of this type, a "proper" rent should be distinguished from a rent on a house or a building, which contains an element of profit or interest, i.e., a

---

[1] Meadows, Denis L. The Limits to Growth. A Report to the Club of Rome Project on the Predicament of Mankind. New York, Universe Press, 1972.

[2] Mihailo Mesarovic and Edward Pestel. "Mankind at the turning point." The Second Report to the Club of Rome.

[3] For example, almost a quarter of the total number of pages in Adam Smith's Wealth of Nations is concentrated in Chapter XI of Book I: Of the Rent of Land.

return to the landlords investment in building the edifice. The rent paid for an edifice is comparable with payments for leasing of capital equipment, e.g. a machine, a lorry, etc. This paper is not concerned with rent for either buildings or machines, but with rents which are or should be charged as costs of resources in scarce supply, for things which are not themselves produced.

II.2 Rents on exhaustible resources

A mine differs from a piece of land in that it is not an indestructible resource. On the contrary, to avoid unnecessary high costs of extraction, it may be desirable to concentrate the production of a certain ore on a relatively low number of mines. All the good quality ore in a particular mine may then be mined in a fairly short span of time, and a new mine is opened when an older mine no longer contains sufficient good-quality ore. The essence of the exploitation of a mine is its eventual depletion, the essence of the good management of a piece of land is that it will remain fertile to produce food for future generations. This difference is, or should be, expressed in the form of rent-payment. Rent on land (cr on any indestructible resource) is a sum of money per month, or per year, while the rent on a mine is in the form of a sum of money per tonne of the ore.

II.3 Why is rent necessary?

Natural resources are a major factor of production. Production methods which are uneconomic in the use of scarce resources should be financially unprofitable. Therefore rent on scarce natural resources ought to be calculated as a cost of production. To make this quite clear, let us consider an example where no factor-rent is calculated: fishing rights.

The increase in the human population has brought us into a situation where the ocean and its food-producing capacity is a scarce resource. If the United Nations Food and Agriculture Organisation had the authority to lease particular areas of sea to certain nations or companies under specific conditions it is conceivable that the following operation would become profitable:

A company would contract the collection of sewage from a certain town, provide a separate drainage system for it, and transport it to its leased part of the ocean. There it would be broken down naturally into minerals and minute particles of organic material, serving as food for small marine creatures, the so-called plankton. The plankton would serve as food for fish and the company would "farm" the area by catching the fish which would be in much greater abundance than elsewhere in mid-ocean areas.

I do not know whether manuring the ocean really is a practical proposition towards increasing the ocean's yield of fish. But if it is a viable proposal from a biological and technical-economic point of view, such farming of the ocean is commercially unprofitable nevertheless. The two main obstacles are:

a) The ocean is free for all and all are free to catch the fish, leaving no profit for the fish-farming company.

b) The price of fish does not cover the cost of operations of this type. If a mid-ocean area would be reserved for the exclusive use of a fish-farming company, this would not by itself make "farming" of that area profitable. "Farming" of a reserved area becomes profitable only after overhunting of the "free" area has reduced the fertility of any free area below that of the reserved area.

If sea areas were leased, the lease would have to be for a number of years, and good fishing grounds near densely populated shores would do higher rents than mid-ocean areas.

This would ensure that operators had the incentive to avoid over-fishing, and the price of fish would cover the cost of "farming" a remote mid-ocean area.

But in fact, fish and other sea animals are hunted in excess of the ocean's capacity to renew at the natural level of food and oxygen supply. Several species of whales are already extinct to all practical and economic purposes, if not biologically as well.

II.4 The Ricardian and the Walrasian Theories of Rent

Economic theory gives us two definitions of the rental value of a natural resource, the Ricardian and the Walrasian. The Ricardian theory assumes that the resource as such is abundantly available, but rents arise because different plots of land or different mines are of different quality. I quote (Ricardo, Principles,[4] Chapter III) - "The return for capital from the poorest mine paying no rent would regulate the rent of all the other more productive mines". To put this in the context of the previous section - the cost of farming a remote mid-ocean sea-area, including the cost of monitoring the fish and providing feedstock for lower forms of ocean-life, would regulate the price of fish and the rent of favourably sited sea-areas with it.

The Walrasian[5] theory on the other hand assumes that the resource itself is in short supply, and its provision at zero cost would cause a demand in excess of

---

[4] Ricardo, David. "Principles of Political Economy and Taxation".

[5] The reference to Walras places the pricing of production factors in the context of the more general theory of economic equilibrium and market-equilibrium by the price. Cf. Walras, L., Elements of Pure Economics, London, Allen and Unwin, 1954. Translated from the French original, Elements d'Economie Politique Pure, (1926 edn).

its total availability. The rent, to be imputed to a resource in scarce supply is the price which will contain its demand to be within the available natural supply. Crude oil is an obvious example. The cost of extracting crude oil from the earth is a fraction of the price.

The following figures are given here on the authority of M. Adelman.[6] Cost-estimate including cost of prospecting, development of field, normal return to capital, etc. (long-run supply-price): ranging from[7] \$0.20 per barrel (Middle East) to \$0.64 per barrel (Venezuela). M.E. actual supply price in 1970 - appr. \$1.20 per barrel (Iran \$1.28, Kuwait \$1.18). Clearly even high-cost Venezuela (and the oil-companies) earned a handsome rent-income even before the 1973 price-jump. Adelman interprets this as an anomaly.

According to the Ricardian model, Adelman's prediction might be correct. But the Ricardian theory assumes the presence of a surplus of poorer specimens of the resource, which are not economically employable even at zero rent. There are coal deposits which are uneconomic to mine and are likely to stay unmined for that reason for generations to come. There is more coal in the earth than mankind can meaningfully use and the deeper deposits and those with the thinner seams of coal will not be mined. The same is not meaningfully true for oil; any known oil deposit of any size is likely to be tapped at some stage during the next 50 years. Hence the Ricardian theory is relevant for coal, although rent is so far not normally charged for mining coal, even from the geologically more advantageous deposits. The Walrasian theory on the other hand is more meaningful in connection with oil.

II.5 Replacement costs

For indestructible resources the rent is the price at which the demand for the service of the resource is contained to the available quantity. As indestructible resources are only available to a certain quantity which is the same in each time period, market equilibrium requires that the rent contains the demand for the resource to the available supply in each time-period. If the rent is too high, part of the resource remains unused, if the rent too low, a part of the demand just cannot be met. This is not so in the case of exhaustible resources.

There is no definite available supply in a particular time-period. If the rent is low, the resource will be wasted and future generations will have to do without it. We should put this in perspective, by realising that depletion of a

---

[6] Adelman, Maurice, The World Petroleum Market, The John Hopkins University Press, Baltimore, U.S.A., 1972.

[7] Ibid., p. 76.

particular resource is not for that reason alone a calamity. The early industrial era has to all practical purposes, consumed the forests which covered large areas of England in pre-industrial times. They have been turned into timber for ships and houses and into charcoal to smelt iron.

In the event, the generation of the Industrial Revolution has compensated later generations, by developing the use of new building materials, e.g., steel, concrete and glass, and new steel-making techniques, i.e. the use of mined coal rather than charcoal. Hence if the present generation will develop alternative sources of energy, before oil runs out, the next generation is compensated. The cost of developing these alternative sources of energy is, of course part of the cost of consuming oil. Hence we will say that a mineral rent covers its stock-replacement if provisions for the development of substitutes are expected to be effective before the resource is exhausted and if the cost of developing these substitutes are covered by the mineral rent.

II.6 The stock containment rent

Let us assume that a research and development programme, aimed at making mankind independent of the availability of some scarce mineral, named X, is under way. The development of alternative techniques, not requiring the use of X, will take time. We obviously have some idea of how long that time is likely to be.

It is then desirable that the demand for mineral X does not in the meantime exhaust the available geological stock. If the rent is to be charged for extracting X from the earth is calculated with a view to this requirement, we will indicate such a rent a stock containment rent.

I would think it likely that something like the 1974 relative price of oil is likely to be adequate as stock containment rent, on the assumption that alternative, environmentally safe methods of energy-production e.g. solar energy, are technically and commercially viable in about 30 or at the most 50 years time.

The level of the stock containment rent will obviously depend on our estimate of the period of time required. If that is a short period of time, the concept is not applicable: the stock replacement rent is the relevant rent. If that time period is somewhat longer the question of the optimal outlay on the research programme may arise.

Would it be worthwhile to speed up the development of alternative techniques, despite higher cost, in order to be able to sustain a high level of consumption of the presently used resource? In principle this question could be ansered in any particular case, given sufficient data by using quantitative methods.

Consider the development of an alternative source of some resource, e.g. nuclear or solar power to replace exhausted oil in either 15 or alternatively 25 years time.

The real income from energy over all the 25 year period is the total energy consumption, less the research expenditure. Society must obviously assess priorities. How much is energy worth, relative to research-outlay, i.e. manpower and other material resources used in the research programme?

Having set a relative price for energy as such, the programme which yields the highest total real income over all of the 25 year period is the preferred one. If the 25-year programme is the preferred one, this also determines the rent on the resource, e.g. oil. It is that level of rent which will ensure that the demand for oil will not exhaust its supply before the end of a 25-year period.

In practice, calculations of an optimal rent is likely to be at best useful illustrations of the problem and the approach to its solution.

There is too much uncertainy about the cost and even about the outcome of research programmes, to be confident that a particular figure and none other is correct. There is also considerable uncertainty about the true magnitude of the geological stock of the various minerals. This uncertainty concerning the geological stock makes the assessment of the stock-containment rent somewhat speculative even if the required time-span is known.

II.7 Stock-Depletion Rent

A resource may be exhaustible, while there is no identifiable prospect of finding an alternative to it. Present technology heavily relies on the use of a wide range of metals. All these metals or their ores are exhaustible resources except in the sense that they can sometimes be partially reclaimed from waste or scrap.

We do not so far have the faintest idea of how to run a technology-orientated society without steel, zinc, aluminum, lead, copper, etc. Hence there is no replacement rent and no stock-containment rent either.

Calculation of rent as a cost-factor in such a situation still means that demand is contained and the resource will last longer than it would if the resource were treated as a free good.

How high the rent should be is then a question of priorities, i.e. our own consumption of (the end products of) these metals or our grandchildren's.

III SOME ENVIRONMENTAL RESTRICTIONS ON THE ECONOMY

III.1 Food

The scope for a further substantial increase in food production by means of the traditional land-based agriculture is, unfortunately rather limited. The main reasons for this pessimistic conclusion can be summarised in two words: fertiliser and pesticides. The problem with fertiliser is largely its limited availability, given the limits on natural resources, e.g. energy and minerals.

There are a few plants and some bacteria which convert nitrogen from the air into minerals on which other plants can grow. A more ancient type of agriculture would therefore maintain the fertility of the soil by planting a "rest-crop" in alternate years, and this "rest-crop" would restore nitrogen to the soil. This practice obviously results in a low yield of food for human consumption, to be cropped from a given area of land, over a number of years. But the production of nitrogen fertilisers by the chemical industry competes with the use of energy as fuel.

Other types of fertiliser, e.g. phosphate, are naturally occurring substances. They are mined by man, put on the fields, and eventually disappear. Some of the minerals will disappear as agricultural run-off into rivers and eventually in the sea, some of it will be absorbed by plants and either be burned as straw or garbage, or return to the soil as rotting plant material. And some of it will be consumed by man as food and ultimately disappear into the ocean as sewage.

But once the mineral deposits are exhausted this support to agriculture is no longer available. The problem concerning pesticides is somewhat different: they become less effective, and are dangerous.

Their purpose is to reserve food, once grown, for human consumption, rather than to be consumed or spoiled by rats, mice, grubs, beetles or fungi.

Unfortunately, pesticides tend to be poison to a wide range of living organisms while this involves obvious dangers to the environment, more particularly relevant to their economic (in)effectiveness is their effect on the natural enemies of pests.

In fact, the more or less random use of a range of pesticides tends to promote pests. I shall come back to this point.

The possibility of obtaining food for an increasing world population from the sea, is, at least in theory, much better. The two main reasons for this qualified optimism are:

a) There is much more sea than land and more than 80% of the sun's energy comes not to land-areas, but to the sea.

b) The organisms, which are the most efficient in converting the sun's energy into carbo hydrates (sugar, starch) are not land-plants but algae, minute plant-organisms living in water. The thermal efficiency of algae is in fact about four times that of rice.

III.2 Pollution

To the general public, pollution is probably best known and despised as a source of nuisance. We dislike the fact that a rush-hour city street stinks from exhaust fumes, and that the water has an awkward taste. Likewise, our technological society allows us to have a holiday at the sea-side, but when we get there we find stains of crude oil on the beaches.

But there is more. There are also health hazards. Air pollution is probably the most widespread form of pollution which is relevant in this respect. Most of the older evidence related to smoke, i.e., particles of soot, ash, and sulphur dioxide. Lead from the exhausts of cars burning high-octane petrol is probably of comparable importance in more recent times.

There are also long-term risks. Man's interference with the environment is on such a vast scale, that it is quite conceivable that certain biological, physical or chemical balances which have so far protected life as we know it, will be upset. Consider the following points:

III.3 Pesticides

For some time, pesticides and in particular insecticides have been remarkably effective in containing damage to crops as well as the spread of certain diseases which are carried by particular species of animals. Effective crop yields have increased, malaria and yellow fever has been contained.

We may hope that pesticides are harmful, only to the particular species which one tries to exterminate. This is, fortunately, more or less the case in some instances, but the bulk of the commonly used pesticides have harmful side-effects.

The most serious environmental danger stems from the so-called organo-chloride compounds (DDT, Aldrin, Dieldrin and related products). These substances are to some degree toxic or harmful to a wide range of organisms.

Organo-chloride are chemically stable, i.e. they are not oxidised in the atmosphere or in the sea and they don't decompose. They are eaten by animals but the capability of living organisms to get rid of them again is very limited. Therefore, many animals (and people), who consume small quantities of them, gradually build up an increasing concentration of these chemicals in their body.

One must assume, therefore, that, for example, the Eskimo people, who live on the fat and meat of animals which themselves prey on fish and other animals, now contain a relatively large quantity of organo-chloride compounds, despite the fact that the concentration of these chemicals in the Arctic Ocean is so far minute.

Furthermore, any chemical of this class, which generally pervades the local environment to which it is applied, becomes progressively less effective as a means to control pests. This is because these chemicals are eventually more effective in destroying carnivorous animals which prey on harmful insects, than in destroying the harmful insects against which they are initially directed.

There is also the possibility of damage to microscopic eco-systems which perform essential biochemical functions.

"Healthy" soil contains a rich variety of organisms and has a certain capacity to absorb nitrogen from the air. If the bacteria and other small organisms are killed by insecticides, the "dead" soil will only sustain the same amount of crop if fed with an increased quantity of fertiliser. By the same token, the capability of the sea, to convert atmospheric carbon-dioxide into oxygen and food for fish is hampered by the presence of pesticides. The same is true for the natural cleaning of rivers below a sewage discharge point.

The position with the more recently used <u>organophosphates</u> (parathion, malathion) is in some respects different. On the one hand these organophosphates <u>are</u> broken down fairly quickly, hence they do not accumulate in the ocean and do not cause such widespread damage to marine life. On the other hand they are less selective. Moderate doses of organochloride compounds do not have any appreciable direct (short-term) toxic effects on mammals and birds, organophosphates are highly toxic for insects as well as for warm-blooded animals.

III.4 Heavy Metals

Metals, like lead, mercury and the more rare cadmium have effects which are somewhat similar to pesticides. However, unlike the organochloride and organophosphate compounds, the substances concerned have always exisetd in nature and we must assume there are mechanisms for their eventual removal from the environment.

III.5 Dust and Carbon-dioxide

Atmospheric dust and carbon-dioxide both affect the climate, but in opposite directions. Dust reflects the sunlight back into space.

The effect of atmospheric dust is amplified by the fact that particles of dust act as condensation nuclii for water droplets, i.e. if man emits dust, smoke, etc., into the atmosphere this leads to the formation of mist, fog and clouds and mist, fog and clouds keep the sunlight from the surface of the earth. Pollution of atmosphere by dust therefore makes a colder earth. Carbon-dioxide on the other hand, has a "greenhouse effect". The incoming sunlight is not significantly hindered by this gas, but the radiation of heat from the surface of the earth back into space is largely in the infra-red range, and this is absorbed by carbon-dioxide. An increase in the carbon-dioxide content of the atmosphere therefore makes for a warmer earth. So far, observation suggests that the effect of the dust is dominant. We don't actually know if this is because of human activity or a natural change in the climate but the earth is getting cooler, at least in the period and area where temperatures were monitored with a view to assessing the overall situation, i.e. since 1965 in the northern hemisphere.

III.6 The Ozone Shield

The rays of the sun as they hit the outer wall of a spacecraft or the surface of the moon, contain a lethal fraction of ultra-violet light. Not only is the proportion of "ordinary" ultra-violet light far in excess of what would be good for any sun-bathing tourist, but a considerable fraction consists of even much sharper invisible light, more comparable to X-rays and beyond that, i.e. so-called V-rays. Living organisms on earth are shielded from this lethal component of the sun's rays by the presence of ozone in the upper layers of the earth's atmosphere. Ozone is a special kind of oxygen and it is derived from normal oxygen under the impact of the above-mentioned lethal component of the sunlight.

But the sun's potential for producing ozone is limited. As ozone is a powerful oxidising agent, it does not occur in any significant quantity in the air near the surface of the earth. It would long have been used up in oxidising some substance rising from the surface of the earth.

It has now been found that the ozone content of the stratosphere is also falling, except for the very outermost and thinnest part of the stratosphere (above 30 km).

The finders of this fact suggest various forms of air pollution which might be responsible for this phenomenon, e.g. atomic bomb-explosions, exhaust fumes of supersonic (high-flying) aircraft, and (by a curious but confirmed catalytic property of these substances) freons, i.e., the propellants used in hair-spray, liquid spray-wax, etc.

III.7 The Nuclear Power Issue

A nuclear power station contains a highly radio-active heat-source, which is used to drive electric generators. The generators themselves are not basically different from those in a conventional electric power station, fuelled by burning oil, coal or gas.

The possible dangers from the use of nuclear energy come under three main headings: the reactor itself, the processing, storage and transport of radio-active material before and after its use in the power-station, and the security risk involved in the possibility that the technology can be put to destructive purposes. The safety of the reactor itself and the safe handling of radio-active material call for technical solutions and safety-precautions like thick walls around potential sources of radiation, etc.

The more authoritative of those who claim to know, contend that it is possible to build safe power stations and processing plants for nuclear fuels. There is a challenge to the received view as well, and the nuclear power programme has been called a "Faustian bargain", implying that mankind is trading its long-term safety for the comforts of modern technology. But even if we accept the scientific and technological claim of the advocates of nuclear power as valid some questions remain.

For a start, common prudence would suggest that our economy should not become dependent on nuclear power until the entire nuclear fuel cycle, from mining the uranium to finding the radio-activity of the final waste-produce insignificant, is based on proper and tested technology.

From that point of view the waste disposal problem is most acute for radio-active materials with half-life in the region of a decade or so.

Radio-active materials with much shorter half-lives, say a month or so have been monitored until not much of it was left. On the other hand, those with much longer half-lives are much less radio-active and minute traces of it, which might get into the environment, wouldn't give rise to a level of ratio-activity which is significant, compared to the "natural" background radio-activity.

So there is a case for waiting, let us say, fifty years to see that no nuclear waste storage-site has leaked, before we place ourselves in the position of being critically dependent on nuclear power.

Concerning the nuclear technology, one danger is that if twenty or thirty countries acquire the capability to build nuclear reactors it is a short step to being able to make atomic bombs. That raises some very awkward questions concerning security.

The human race does not have a particularly good record concerning violence between different groups within the species. More in particular, the recent historical record has it that it is highly likely that at least one of any group of twenty or thirty states will turn into a mad or criminal regime at some stage in the next quarter-century. What would the world have looked like if the nuclear technology had been wide-spread at the time of, for example, the coming to power of Hitler in Germany?

A related problem in this context is the presence of radio-active waste-dumps. They may, we hope, be safe if they are left undisturbed. But what if they are used on purpose as radio-active poison?

IV THE ECONOMY AND THE ENVIRONMENT

If we accept the submission that environmental factors restrict, or ought to restrict economic activities, the question rises as to what to do about it. To begin with, certain activities, e.g. the sale of certain kinds of hair-sprays, may be expendable, but are nevertheless known or suspected to be dangerous. They ought to be banned. But the large-scale use of various fuels and metals is basic to our technological society. We have burnt coal and oil, smelted steel, zinc and aluminum, build ships and aircraft and so far the next ice age has not started. It becomes a question of containing the unbridled increase rather than banning the discharge of the economy's debris into the environment.

Resource-pricing economics would then have it that each environmental absorption limit should have its own price. Firm's action in choosing the most renumerative methods of production would then lead to the maximum economic product which mankind can afford within the set limits of environmental absorption capacity.

Unfortunately, we do not always know the tolerance limits of the environment. If we continue to pour debris into the environment, up to the limits where we think things will become dangerous, we are likely to land ourselves with problems we had not thought of. Our ignorance may in some cases make us more prudent than is strictly necessary, but this will be little consolation if we get it wrong with something else.

We therefore need a guideline of general prudence, in addition to specific restrictions on identified forms of pollution.

The guideline which is proposed here is that the extraction of all raw ores from the earth, whether geologically scarce or not, should be restricted to substantially its present level. As all smoke, effluent, etc., is obtained by processing of raw materials into something else, this would automatically imply some degree of containment of pollution. The containment of the demand for raw materials would in this way stand "proxy" for the protection of the environment, additionally to direct controls or changes on harmful activities.

Note that the containment of the use of raw materials is not the same as zero growth of production. A raw material restriction whether real in its own right or imposed in order to protect the environment obviously implies a reduced rate of growth of production. Relative to what would be possible if there was no curb on the supply of raw materials we could have less production in obsolute terms, but more comfort per tonne of material used.

This would be attained by charging a mineral rent, a stock depletion rent on all minerals extracted from the earth, irrespective of whether they are likely to run out in the next fifty years or so. The stock depletion rent would be aimed at stabilizing the consumption of raw materials, rather than at preserving the geological stock for a specific time-period.

The increased amounts of comfort per tonne of material would then come from the "normal" incentive mechanism of the price-structures. More efficient isolation of houses, increased thermal efficiency of mechanical equipment and more economical use of power as such, would all be stimulated by the high cost of energy. High costs of newly mined materials also stimulates the recovery of similar material from waste or scrap.

Or, even better, higher costs of raw materials will make the repairing and re-fitting of defective equipment economical, in preference to replacement by new equipment. New fabrication benefits from the possibility to standardize, mechanize and automate large-scale series production. For this reason, fabrication from raw materials often requires less human labour than the repairs of a previously fabricated but now defective item of the same article. Diagnosis of the defect and replacement of a few faulty components in a defective item of equipment cannot be done in the same standardized way as new fabrication. The decision to repair or to discard and replace by a brand-new item of similar equipment therefore often is a choice between the use of human labour in repairing, or the use of raw materials in new fabrication. Another issue is economics of scale and the trend towards specialization of industry, combined with concentration in certain urban and metropolitan areas. The construction of big specialized factories is often efficient in terms of labour-cost, sometimes also in terms of materials consumed in situ per unit of output. But it also involves transport over long distances.

A big factory requires a wide market. As fuel to run a transport system becomes costly relative to human labour, the efficiency in terms of labour-productivity has to be balanced against higher transport cost. High costs of fuel therefore mean that in those cases where the material requirements of production, i.e. raw materials are available over a wide geographical area, smaller production units become more economical. The same is true in those cases where the raw materials are less bulky (or more amenable to storage and bulk-transport) than the end product. For example, bread baked in Inverness is currently transported over large areas of the Scottish Highlands and even to the Hebrides Islands on the Atlantic side of Scotland. Clearly fuel for lorries and ferries to the islands would be saved if the wheat were transported instead.

To tax the extraction of minerals from the earth stimulates the use of human labour in preference to raw materials. The heavy industries involved in the processing of these raw materials are major sources of environmental pollution, therefore the more economical use of raw materials also contains pollution. If mineral rents are charged to protect the environment, they should hit the mining of geologically abundant materials as well as scarce ones. Discrimination between one material and another should be based on relative harmfulness rather than on scarcity. The 1974 four-fold jump in oil prices will no doubt help to conserve Middle Eastern oil stocks. But from an environmental point of view it is containing the use of energy as such, not just containing the use of oil which matters. To the extent that dearer oil stimulates the use of coal this is no advantage from an environmental point of view: coal burns more "dirty" than oil. If higher energy prices are to contain air pollution they should include higher coal prices, even where coal is not nearly as scarce as oil. Similarly, the one-sided increase in oil prices has made nuclear power commercially viable, and disposal of radioactive waste-products in the environment with it.

V THE RAW MATERIALS AND CURRENCY CRISIS

Between August 1971, when the U.S.A. suspended the convertibility of the dollar into gold, and October 1973, when the Middle Eastern oil producers announced their production cutbacks, the world economy made the transition from apparent stability to rampant inflation and uncertainty about the future.

This section is written in the conviction that things will not be as before anymore, and that we shall not understand the problems which now present themselves, if we do not first understand why the economic storm broke out at that particular time in the first place.

The increases in the prices of a wide range of raw materials which suddenly occurred in 1973/74, were triggered off by a combination of the currency crisis with a shift in the demand-supply balance in the direction of shortages. For two and a half decades industrial countries had been busy to increase the stock of machines and the supply of technologically trained manpower. It should therefore be no surprise that raw materials to fuel these machines and to be processed by them, rather than machines and the ability to operate them, are now in scarce supply. But the actual price-development of the 'fifties and 'sixties was in the direction of higher wages, dearer machines and cheaper raw materials.

I have discussed this anomalous development in the 'fifties and 'sixties in some detail in the eighth chapter of my book.[8]

I shall now pay some attention to the causes of the sudden reversal at this particular point in time, and the likely short-term results. I would say that the development of the internal economies of the leading industrial countries during the later part of this period, i.e. the 'sixties, was one of financed cost-push. There was a moderate degree of inflation, but one could not fairly describe that as demand inflation. The two main factors for which demand did rise faster than potential supply, were mined raw materials and environmental resources. The prices of oil, coal metal-ores etc. were more or less stable, and environmental resources had no price at all. But there was a systematic disequilibrium in the way the increased value of trade was financed. The system of fixed parities as set up after World War II, was essentially a gold-dollar system.

The stock of gold increased at a much slower rate than the value of world trade. The stock of internationally acceptable money, i.e. gold and other countries' holdings of gold-convertible or dollar-convertible currencies of leading industrial countries increased at a somewhat faster rate. This was mainly because the United States had a deficit on her balance-of-payments and other countries' holdership of dollars increased. This financing of the supply of "world money" by the US could not be continued indefinitely. The number of dollars increased, while the stock of gold needed to guarantee the dollars' gold-convertibility remained virtually stationary. Already in the mid-fifties the US could not in fact convert all these dollars into gold, should that have been asked. From 1954 onwards there were more gold-convertible dollars in the rest of the world than gold in the USA.

---

[8] A. R. G. Heesterman, 1974. Macroeconomic Market Regulation. London, Heinemann, 1974.

During the 'sixties the gold-dollar ratio continued to decline. But at the same time the ratio between the total stock of internationally held money (gold plus foreign exchange) dropped from 11 month import-reserve in 1950, via 7 months import-reserve in 1960, to 3 months import-reserve in 1970. Then, on 15 August 1971, the US suspended the convertibility of the dollar. The dollar era, i.e. the period in which the economy of the rest of the world was regulated and to some extent stabilized by the economy of the United States, came to an end on that day. It was followed by a period of uncertainty. In 1972 and 1973, things went on still much as normal in the material "real" economy, but exchange-rates as well as commodity prices were almost certainly affected by speculative buying of relative "hard" currencies like Deutsche-mark or Swiss-francs, as well as of raw materials. Whether speculative purchases were by bona fide users who bought earlier than they would otherwise have done, or big full-scale speculators who intended to re-sell the same material at a profit, is in this respect less essential than the mere fact that materials were bought for speculative purposes.

One may interpret the flight from money into goods as an example of Gresham's[9] Law:

> Bad money drives out good money. Raw materials become non-available to (other) bona-fide purchasers, not because they are really more scarce, but because they were hoarded as a store of wealth by people who distrust other forms of storing wealth.

After August 1971, the dollar was seen to be vulnerable to both inflation and loss of external parity, and the owners of big sums of dollars wanted to off-load them. They bought either Deutsche-mark or Swiss-francs etc., relatively stable currencies, or else raw materials. Should this speculative demand for raw materials have arisen earlier (to some extent it did in the 1950-51 boom associated with the Korean War), it might have been met out of available stocks and increased production. In that case, speculative buying would not have been activated further by sharp actual price-rises creating the expectation of more inflation to come. Instead the currency-crisis coincided with a strengthening of the market-position of the sellers of, in particular, low-cost protein, cereals and crude oil.

---

[9] Gresham's Law states that when the government mints coins with a lower content of precious metal than was formerly the case and issues them for the same legal tender value, the "good" oil coins will not circulate side-by-side with the "bad" new ones. The "good" ones will be hoarded, melted down or exported and only the "bad" new ones will circulate.

This "law" is ascribed to Sir Thomas Gresham (1514-1579), but the historical correctness of Sir Thomas' authorship of this "law" is debatable. The term "Gresham's Law" is, however, well established in the tradition of economic theory.

For the agricultural commodities, the reasons for this shift in the direction of shortages was to some extent incidental, i.e. an unexpected disappearance of the fish from the sea along the coast off South America, and harvest mishaps in some countries, notably the Soviet Union.

For crude oil, it was a case of deliberate manipulation of the market. Whether the production cuts decided by the Middle East producer states in October 1973 were initially meant as an economic rather than a political action, is a matter we can speculate on. The fact is that if the October 1973 action, of the Middle East producer states is judged by its results, the war with Israel helped the Arab states to achieve the unity needed for a collective action.

In pure economic terms, the action taken amounted to the creation of a cartel, which drove up the prices by cutting back on production. The energy costs of fertilizer and tractor-power then obviously affected agriculture as well as industry and transport.

VI SOME INCOME-EFFECTS OF THE RAW MATERIALS CRISIS

Price-changes, as experienced in 1973/74 obviously resulted in a pronounced change in the distribution of income between nations. The income-effects may be summarized as follows.

Three groups of countries enjoy increased real-incomes as a result of higher prices of their exports. The first group consists of the producers of oil and some related materials, natural gas, etc. The general characteristic of this category of materials is the fact that their extraction is capital intensive, hence large amounts can be extracted even in countries and areas where very few people live. Oil is presently the one major resource coming in this category. But if my suggestion to use raw materials as a proxy for the protection of the environment is followed, the similar problem will arise with other minerals as well.

The combination of a sharp rise in price, with a high rate of per capita production has meant very marked increases of income per inhabitant. For example, in 1971 Kuwait exported oil to the tune of about 3,700 million US dollars, on a population of only about 500,000,all women, children and non-Kuwait servants included. That amounted to about five thousand dollars per head. For 1974, one must assume that, despite the cuts of October-December 1973, production is still well above the 1971 level. With the prices risen to about five times its former level, oil-exports now are at least $25,000 per head. Not all of this is pure income for the Kuwaiti, the oil companies also take their share. But the distribution also shifted in favour of the Kuwaiti government.

While Kuwaiti is an extreme example, per capita oil revenue in the several million inhabitant countries (Iraq, Libya and Saudi Arabia) was in the order of $200 to $1,000 per head in 1971 and must also be about 5 times that sort of figure now.

The second group of beneficiaries are those of the relatively densely populated low-income countries, who happen to be lucky enough to be endowed with scarce and also well-developed natural resources. They have experienced sharp increases in their export-revenues, on account of higher prices of some of their major export products - unlike the small oil-producing countries, who cannot possibly spend their suddenly increased income, countries like Indonesia (112 million people), Nigeria (62.6 million people), Brazil (86.2 million people) have no trouble to spend their increased income. Per capita income figures are not particularly striking, and imports are rising fast. The raw materials concerned are sometimes agricultural (cotton, rubber, sugar), and in the case of Indonesia and Nigeria also oil, besides other raw materials.

The third group of beneficiaries are some of the high-income advanced countries which benefit for the same reasons (Canada, Australia). Australia had, oddly enough, a balance-of-payments deficit in the first six months of 1974 despite increases in the prices of its two major export products (wool and wheat) which were hardly less spectacular than in the case of oil.

Two major groups of countries are poorer in real terms than they were before the big price-jump. The hardest-hit group is the densely populated low-income countries without "strong" export products. They depend on import of food and raw materials and they cannot pay these imports at their new, higher prices. The problem may be illustrated by comparing the typical example of this group (Bangladesh) with another densely populated low income country: Indonesia. Both countries have an overstrained agricultural sector, which at its present state of development just cannot feed the people. Should a harvest mishap strike Indonesia, this will cause severe social dislocation: peasants will flock to the already swollen cities. But come to the crunch, Indonesia can afford to feed the teeming millions of Jakarta and other big cities, by paying imported food with hard cash from the export of oil and other scarce minerals. Bangladesh cannot, and people die from hunger.

Densely populated industrial countries, e.g. Japan and Western Europe. Superficially, it would appear that this group is almost as hard hit by the raw materials crisis as the poor countries. Export prices of finished industrial products have not increased by anything like the equivalent of the price-rises of raw materials.

But this group of countries is shielded from the full impact of the raw materials prices by the fact that they are at least up to now allowed - at a price - to pay for their imports by issuing their own money.

Should the day come when Kuwait would sell oil, only against payment in Kuwaiti dinars, Canada sell wheat only against payment in Canadian dollars, it would certainly mean unemployment and poverty, if not hunger in Brussels, London and Tokyo. One may sincerely hope that that day will not come, but that is just what the rest of the world is doing to Bangladesh.

The income-effects of the increases in raw material prices are not limited to changes in the distribution. The total is reduced.

This is so, not only because the supply of raw materials is reduced, but also because production of industrial and other non-scarce products, including agricultural products, is hit by dislocation of international trade in general.

The income-benefits to some countries with small population are so huge and so sudden, that they cannot possibly spend their increased income. For oil alone, I would estimate the income of the six main exporters with relatively small populations, to be about a third of the total stock of internationally acceptable money. Hence, if the exporters do not spend their money all of the present reserves would be in the hands of the oil exporters in three years' time. Unfortunately, the oil countries were already piling up big surpluses <u>before</u> the oil crisis. They just do not have the people to spend it, and lack the facilities to transport and use imported products.

The relation between population, per capita export-revenue and import fraction is illustrated by the following tabulation, referring to 1971.

Country	Population (millions)	Export-revenue per head ($)	Import/Export ratio (%)
Iran	27	100	80
Iraq	9	170	53
Venezuela	10	300	72
Saudi Arabia	7	600	21
Libya	2	1,400	26
Kuwait	0.5	5,400	28

VII THE RECYCLING ISSUE

The oil exporting countries are accumulating and are likely to continue to accumulate substantial surpluses of export-earnings over the costs of their imports, for the foreseeable future. Or they should, if we consider the oil-price as a curb on the use of energy in general. These export-earnings must come from somewhere, and the rest of the world will be in deficit. The question arises as to what financial form this deficit will take.

First of all the producer countries may accept payment in U.S. dollars, English pounds, German marks, etc. That appears to have been the case, to the tune of about $10 billion, in the first half of 1974. This figure is less than 20 per cent of the balance of payments surplus of these countries. Obviously, a sizeable part of the surplus took other forms than money-type claims on the central banks of leading industrial countries.

The following alternatives are possible in this respect:

Medium-term credits to governments and central banks of leading industrial countries. Press publications indicate that has indeed happened on a considerable scale.

Purchase of bonds, shares and real estate in the consumer countries, via the normal internal markets for these assets in the consumer countries, i.e. commercial banks, the stock exchange and estate agents.

Direct participation in industrial ventures in industrial consumer countries.

Ditto in Third World countries.

Food and general income-aid to poor countries, i.e. financing other countries' deficits on their imports either food, and industrial consumer goods.

Of these various alternatives, only the last one has the outward appearance of a free grant-in-aid. In fact, <u>all</u> forms of "recycling" of oil money are likely to contain in substance an element of grant-in-aid.

The reality is that re-cycled oil money finances not additional investment, but continuation of import of oil. The rest of the world cannot genuinely pay four time the price per barrel as it used to do, unless one of the following conditions, or a combination of a "mixture" of them is met.

<u>Oil-countries buy more</u>

The rest of the world can pay, by supplying four times as many television sets, motor cars, sugar, new factories, etc. While this cannot necessarily be excluded to be the case in say, ten to fifteen years' time, this is precisely what so far has not been the tendency. Their 1974 earnings-torrent has indeed prompted the oil-producers to go on a shopping expedition, but it doesn't look anything like enough.

More inflation

The rest of the world can experience so much inflation that it ends up with the changing also four times the former price for its exports to the oil-exporting countries. That would make the whole operation rather meaningless.

A "slump"

World trade can contract to a quarter of its former level, and industrial production, transport, etc. with it, causing a four-fold drop inthe consumption of oil. One would sincerely hope this was not the "solution" by which this problem will eventually be solved.

Re-allocation of resources

There could come a drastic shift in the composition of the demand for energy, away from oil, or at least away from oil from these countries. The development of less energy-intensive processes would have the same effect with respect to oil-consumption.

Of these alternative possibilities the last one, the "re-allocation" solution, is perhaps desirable and indeed the claimed rationale of the operation. But it is clear that this adjustment will require time, and that substantial sums of money will need to be re-cycled. In the meantime, the "slump" alternative would become reality. That "slump" will obviously not be of the severity indicated by the factor four, some part of the gap will be absorbed by increased imports of the oil countries and by inflation.

VIII THE DISTRIBUTION OF FOOD

Should the issue of re-cycling (or to put it more honestly: redistributing) rents be with us for any length of time, or re-appear with other resources, we must consider a serious complication. We do not only need to re-distribute money, but also food, even if that is not the food of the areas with surplus-income.

If the trend towards a cooler* climate continues or if there is a major pest-control failure, there will be only two ways to balance the global food-account: by starvation or by the rich eating less, in particular less meat and drinking less wine. Aid to help poor countries with importing fertilizer and with building irrigation structures, wells, etc. won't be enough. But, although people can live without them, meat and wine are socially considered basic requirements in

---

* This passage is in effect a post-conference insertion, i.e. was written after the June 1975 mini-winter in Northern Europe.

the regions of their highest consumption.

Yet a hungry world cannot afford to feed edible food to cattle and pigs, certainly not to the extent that this has been done recently. Likewise the Mediterranean area should grow food for Northern Europe, preferably several harvests per year on irrigated land. But to make this profitable for farmers while at the same time driving significant quantities of meat and wine out of the budgets of the affluent, would require a further steep rise in the price of food generally.

That is, to drive meat and wine out of the full-employment income budget of the affluent but densely populated developed regions, would require a rather stiff increase in the price.

A moderate raw materials crisis without re-cycling of the rental income might well do it, at the cost of widespread unemployment of labour.

Balance of payments problems would cut the import of animal fodder directly, and the industrial unemployment would simply reduce income. Unemployed people eat less meat and drink less wine.

Because only Northern Europe and Japan, both relatively affluent regions are large-scale food importers, a high price of food and/or a moderate recession in world trade are both congenial to a more equal distribution of the available food supply. Once a world recession is there, incidental imports needed because of harvest mishaps in particular countries, could be made available from surplus stocks of the wealthy countries.

Thus the success of recycling rent from one source, e.g. oil, would prevent the appearance of food-surpluses and the inequities in the full-employment income distribution would have much harsher results.

Raising the price of food would bring back the surplus-market selectively for food and not for human labour. But that would give rise to further complications. The issue of the re-cycling of land-rents would arise. And one would have to contain too large inroads into the remaining areas of forest, moorland, etc. as well as for example, overfishing of the sea.

IX SOME FURTHER DISEQUILIBRIUM EFFECTS OF INFLATION

The following additional complications arise in connection with the rampant inflation which has developed in the wake of the raw material crisis. Firstly, there is the possibility, indeed the logical development, of a flight from money into goods. In concrete practical terms this means that it is quite possible, indeed likely, that stocks are held for purely speculative purposes. It could be that food is being stocked in the wealthier consumer countries, not because it is currently needed there, but because it offers the prospect of profit (especially if the stock is financed by means of a credit), while somewhere else on the world people are dying from hunger.

Next, there is the problem of erosion of agricultural guarantee-prices. There is aphenomenon called the "pig cycle" which is well known from economics text books. When the price of some agricultural products, e.g. pig-meat is high, farmers start to make it. But if they all do that, there is a glut of pork by the time the pigs have grown. Then the bottom drops out of the market and everybody moves out of pig-breeding. Whereupon pork becomes again very expensive. For storable agricultural products, e.g. cereals and indeed for meat, this problem has largely disappeared in the post World War II period, as a result of the system of agricultural guarantee-prices, coupled with publicly financed intervention stocks.

That system assumes that when the officials of the Department of Agriculture fix a guarantee price, farmers can be sure that by the time they get their money on the basis of this guaranteed price, the guaranteed price allows them to buy seeds for the next year, pay their bills and still have enough money left to support their families. Unfortunately, a guaranteed minimum price becomes meaningless, when costs of production escalate in rampant inflation.

Already there is evidence of a "beef cycle" which has developed on lines not unlike the economics textbook has it for pigs. Only, prices do not drop, they just stay semi-stationary and production becomes uneconomic because of sky-rocketing costs.

The absence of a meaningful price-guarantee also means that the incentive response of the prevailing price is weakened. The sharp increases in the prices of staple food products in 1973 ought to have provoked a sharp increase in the supply of these food crops, but for the fact that no-one knew whether similar prices, let alone similar relative prices, would prevail again in the next year.

The combined result of speculative stock-building and erosion of guarantee prices is likely to be greater uncertainty about future prices. Relative prices may have started to re-adjust in a desirable direction, but they may, for all we know, well have overshot the long-term equilibrium position and the whole story may need to be re-told, with the opposite sign, on account of depressed raw material prices.

X SOME INSTITUTIONAL ISSUES

The central question of economics is how to make the most efficient use of scarce resources. Therefore, whenever particular customs and institutions inhibit the efficient use of scarce resources, economics becomes a normative social science. I submit that the division of the world in nation-states and the absence of a body with clear cut authority to restrict and regulate the use of the earth's resources impeded the efficient use of these resources.

The second reason for an economist to make statements of a social-political nature is given by the fact that the relative scarcity of resources influences the distribution of income between different social groups. Here the norm is not the economists' but society's. The economist 'merely' applies existing social-political norms and values. These norms may be either generally accepted or controversial. In the latter case the economist will take a position not as economist but as citizen.

But it is the economist's duty to tell other people when changes in the relative scarcity of resources are likely to give rise to new distribution problems or to give more urgency to existing unsolved problems of income distribution.

I submit that the transition from a labour-supply contained economy to a material resource-contained economy gives rise to a new problem of income-distribution. The concepts of equity and solidarity between people belonging to different races and nations require transfers of income between nations. Alternatively some natural resources could be considered as belonging to mankind as a whole rather than to any group or nation in particular. Hitherto most of the income from production has been imputed to labour, and there was some degree of 'natural' equality in income. Every person owns his own labour. There is no "natural" reason why minerals, solar energy or clean ocean water should bear any relation to numbers of people in particular countries. A greater emphasis on natural resources as a cost of production may be necessary in order to contain the demands on these natural resources but it inevitably gives rise to new problems of income distribution, which can only be solved in the direction of a greater degree of equality by international transfers of income, or by exploitation of these resources by supernational organisations in the first place.

In theory, we could consider a containment of the demand on natural resources without equality in the distribution of the income. Such a "market mechanism" solution obviously concentrated on the conservation of resources for which there are identifiable owners and sellers. Raw materials like oil and metal-ores rather than environmental factors would earn the rental income. The associated price structure would stimulate the development of nuclear power, the burning of coal etc., whether environmentally desirable or not. Restriction of the demand on non-owned resources like the atmosphere or the oceans presupposes a criterion of justice as to who will be allowed to have a certain quantity. I submit that that criterion can only be found in the definition of a just income distribution.

I do not think it is very likely that the logical solution to these two problems, the creation of a world wide economic authority will become a reality in the near future. It seems more likely that mankind will not restrict itself to a rational use of the earth's resources. Unless we stumble into a world slump, some resources will be destroyed in an unregulated free-for-all and a disproportionate share will accrue to the more greedy among those who in fact (over) exploit them. (The whales are now practically extinct, because the whale-catchers couldn't agree on who was not to catch them.) The pessimistic assessment of our social-political reality is no good reason for not spelling out how things could be done, if mankind wanted a rational use of its resources and wanted the revenue to be shared out more equally than is presently the case.

The functions of the world economic authority ought to be:

To make reasonable plans concerning the demand for important natural resources. These plans should be consistent with one another within the requirements of known or projected technology, while being within the overall norm of containing the demands on natural resources at an acceptable, non-increasing level.

To recommend prices and quota which may be expected to keep the demands on these resources within the projected limits and to take appropriate steps towards the regulation of international trade to enable the world to realise the highest level of economic activity consistent with the budgeted use of natural resources.

To mediate between countries and where necessary to impose levies and grant subsidies, in order to ensure a distribution of world income which is sufficiently equal to guarantee basic material security, i.e. non-starvation for all the world's citizens.

To encourage, and where necessary to supplement by its own actions the production of basic food requirements, irrespective of the commercial profitability of such operations. In particular, the application of science-based capital-intensive methods of converting marine resources into food does not appear to have much commercial prospects, except possibly in Japan. In the wealthy countries the need for additional food is not urgent and the poor countries lack the technological capability. An international agency would appear to be a solution, at least to coordinate and finance research in this field.

X.1 The World's Monetary System

We must re-establish financial stability in one form or another. Rates of inflation of 20 per cent per year or more make reference to sums of money to be paid at some future date, several years in the future, almost meaningless. Pensions become worthless in a few years time. Sums of money put in a budget to allocate funds to various departments within large organisations become meaningless. Calculations of returns to investment become virtually impossible, taxes cease to refer to genuine income as a result of artificial book-profits on stocks and work in progress. In short, inflation at rates above, what can be called "creeping" inflation makes rational calculation in economics impossible except by reference to some other standard than the actually circulating kind of money.

The problem of re-cycling oil-money and any other similar windfall-rents which may rise comes into the same category. Inflation disturbs the cost of borrowing and the return to lending. Re-cycling oil-money is not a question of investment it means financing consumption. The countries who cannot otherwise pay the higher oil prices live on borrowed money. If interest in non-inflating money were to be paid on oil-deficit loans, it would become clear that this is something which the consumer countries cannot afford to do for any length of time. Either we ought to call it an income-transfer, or we should consider oil-loans as a strictly temporary device, to cushion the suddenness of the price-change. The long-term solution would in that case be a drastic reduction in the quantity of oil consumption. The present rate of inflation amounts to an attempt to keep the true nature of the problem hidden and this can only lead to the creation of ill-will. The creditors of the oil-producers will feel cheated out of a proper repayment of their loans.

I submit that asking for outright grants is less likely to spoil international relations than to ask that oil consumption be financed on credit while offering repayment in rapidly inflating money. The element of "bad faith" in the relations of the industrial consumer countries with the oil countries is enhanced by the fact that the consumer countries appeal to the producer countries' "duty" to co-operate in keeping the world economy going. As the inflation is world-wide, a world-wide solution is called for. It is, in this connection, a great pity that the Dutch-South African Professor Jan Goudriaan published most of his writings in Dutch rather than in the English language. Professor Goudriaan has argued for many years that a link between money and an agreed combination of raw materials would stabilize both money and the prices of raw materials.

The original Goudriaan-plan apparently envisages the stabilization of a largish number of prices of individual raw materials. That is probably a weakness because a realistic system ought to accommodate the possibility of changes in relative prices. The Goudriaan plan bears in this respect the stamp of its association with the depression of the 'thirties'.

Instead, I propose that I.M.F. drawing rights or any succeeding form of world money be index-linked. This would in practice mean that an announcement to the effect of something like the following would be made:

"The IMF price-index consisting of the following batch of basic raw materials---------

(read a list of commodities, their observed prices in a base-period, converted into IMF drawing rights, and their weights in the index)--------

shall be 1.00 Transactions with I.M.F. shall be in constant-prices Drawing Rights"-

It would then be up to the I.M.F. to adjust national currency's exchange rates in consultation with the governments concerned. Should no agreement result from such consultations, I.M.F. could ask the country in question to settle its financial transactions vis-a-vis the I.M.F., by payment in kind, by delivery of units of the quoted batch of raw materials. One would need to organise some acceptance and delivery points, and maintain a certain warehouse-capacity, but I do not envisage that intervention stocks should play an important role in ensuring monetary stability.

For a start, just now is not a very practical time to build up intervention stocks. Raw materials are in short supply.

Secondly, the intervention stocks of mineral raw materials are in fact geological stocks in the ground. The issue of Special Drawing Rights or any other form of credit or grant would amount to a licence to have some of the available geological stock mined. For this reason it is obviously essential that a scheme of this kind should be run in co-operation with the main producer countries of essential raw materials, in respect to oil, that is a reality to which the world will have to adapt itself anyway.

X.2 Internal taxes and resource-rents

At the relative prices of human labour and natural resources, which prevailed in the late sixties it is likely that containment of the demands on natural resources is possible, only at the cost of widespread unemployment. Yet, there is no question of an absolute surplus of human labour.

I suggest the following measures to stimulate useful employment of human labour, while containing the demand for natural resources.

First of all, taxation should be on the exploitation of material resources rather than on income. I propose that governments impose taxes on the extraction of various minerals from the earth. The activities of publicly-run bodies like the National Gas Corporation, Coal Board, etc. should be taxable on the same lines as similar activities by private companies. Effluent and smoke charges, as far as practicable from the point of view of monitoring come in the same category and should also apply to all discharges whether by private firm, public corporations or local authority. Such taxes would obviously raise production costs, and prices of gas, electricity, steel, bricks, gravel, cement, etc., would go up. On the other hand, taxes on personal income would not necessarily be needed as a basic source of public revenue.

They would eventually be phased out, with a possible exception of "surtaxes" on the very high incomes. There would then be no need to increase gross earnings from employment. This would also make it easier to re-direct public spending towards things like health, education, care of the elderly rather than outlay on material-intensive investment. Or, more precisely, the same public budget as at present, would in fact mean more education and health services, and fewer motorways and aircraft.

The transition from the one system of taxation to the other would have profound effects on the relative costs of various products. To avoid serious dislocation of existing social and economic structures it would therefore be desirable to introduce the new system of taxation only gradually.

X.3 Control of other prices

The prices of exhaustible minerals and other exhaustible natural resources, relative to human labour and industrial products are indeterminate in the short run as well as in the medium term.

Raw materials prices have recently shot upward, we traced the cause to a combination of a stronger long-run market position, speculative buying and manipulation of the market. The proposal to monetarise raw material prices in terms of IMF drawing rights would (if accepted) largely stabilise raw materials' prices in absolute terms. The relative price in terms of human labour and industrial end products would still be indeterminate. The indeterminateness in absolute terms would be restricted to wages, prices of industrial products and national exchange rates.

I propose that these be made determinants by means of the following arrangements.

a) Fixed exchange rates to be quoted relative to the IMF drawing rights.

b) The IMF should undertake to support national currencies at the official rates, on the condition that the internal economic policy of the country in question, was consistent with guidelines of world economic policy. The containment of domestic costs would be a major point in such guidelines of world economic policy.

X.4 Better give away what you can't do anything with

I now have to come back to the re-cycling issue. The system of price stabilization, proposed in the previous section, would appear largely to settle the questions as to how prices which do not figure in the IMF drawing rights price-index could be stabilized. It leaves the question as to what norms should determine the level of prices which the envisaged guidelines of world economic policy would aim at. Our general guideline of containing the demand on natural resources suggests that this should be at a fairly low relative price of labour. The implied income-distribution, with its high rents on natural resources would be a more unequal one than we have known in the second and third quarters of the 20th century.

It is factually possible, but undesirable for the world to return to the situation of the sixties, to depletion of the existing resources at a fairly high rate and to low prices for those raw materials. On the other hand, if the higher relative prices of raw materials are there to stay, we shall have to face the question of the transfer of the rentier-income. Rentier countries ought to spend and as far as possible should be pressed to spend, a sizeable fraction of their surplus on the development of more populated, poorer countries and on outright income-aid to poor countries. This may sound an unrealistic appeal to philanthropy, but consider the alternatives:

The surpluses could be spent on local development of the rentier countries themselves and on their own consumption. As they are already above the starvation level, the increase of their consumption would be largely on services (import of doctors, teachers and technical experts from outside) and on industrial products.

At 1974 relative prices, the per capita accounted income of the typical 5 to 10 million inhabitants of oil countries is still lower than in the typical Western European countries. But this assumes continuation of the present production-structure of these countries. Should they become industrialised their income would be considerably higher. The sums of money are large enough to build something like another New York and another Tokyo in a single generation. In that case the consumption of the scarce natural resources is shifted to the home-countries of their owners, and the consumption of the corresponding environmental emissions, exhaust fumes and the like, with it, not to mention the drastic social transformation.

Meanwhile, one would have to accept the death by starvation of quite large numbers of people in densely populated poor countries, due to lack of fertilizer and lack of mechanical transport to bring food to people in areas with poor harvests. Poor countries get poorer as a result of the higher cost of oil. All oil-importing countries get poorer. Alternatively, the surpluses could be "invested" in various forms in industrially developed countries. This is what so far is keeping 1974 trade more or less flowing.

If the higher prices of raw materials succeed in their basic aim of containing the demand for these materials, the rentier countries will not need a repayment until well into the 21st century. The industrial consumer countries, on the other hand, cannot repay, except by effectuating that painful reduction in their consumption, which they are so far avoiding by means of borrowing.

If the day comes when the oil is indeed finished, oil-consuming countries are hardly likely to be able to do then, what they cannot do now. There is no need for additional investment in industrial countries at a time when industry is slowing down because of scarcity of raw materials, and the "investment" of oil-money is a disguise for the financing of consumption in industrial countries. In short, "investment" as a means of re-cycling oil money is throwing good money after bad money anyhow. The producer countries might as well try to get some political benefits from being generous now.

One side-result would be that the oil-consuming industrial countries would be in a position to earn money needed to pay for at least some import of the dearer raw materials by selling to low-income countries, instead of having to borrow money which they cannot pay back.

COMMODITY MODELING APPROACHES TO RESOURCES, ENERGY, AND REGIONAL PLANNING

Walter C. Labys

I INTRODUCTION

Problems of resources, energy and regional planning need to be solved within a strong analytical framework. This has given rise to a number of modeling approaches or methodologies which can analyze problems, test policies and make forecasts of future outcomes. For example, a number of models have been constructed which explain the extraction and consumption of commodities of a natural resource type. This would pertain not only to deep mining and strip mining but also to the use of the ocean floors. Other models relate to petroleum and energy commodities. While yet another group relates to the impact of commodities on regional development. The purpose of this paper is to survey the methodologies which have been used in constructing such models, particularly those that refer to commodity market analysis. After a brief review of the basic methodologies, their application is then explored in terms of some basic areas of interest namely, resource models, energy models and regional planning models. In each case, suggestions are given for future research.

II METHODOLOGIES

By a commodity model is meant a quantitative representation of a commodity market or industry, where the empirical relationships included reflect the underlying demand and supply conditions as well as other economic, political and social phenomena. The structure of such models normally derives from micro-economic theory in which demand, supply and inventories interrelate to produce a market price. However, as shown in Table 1, the structure finally adopted depends on a number of considerations such as the nature of the methodological approach, the quantitative techniques of interest, and the economic behaviour specified. The most elementary methodology is a <u>market model</u> represented as follows:

$$\begin{aligned} D &= d(D_{-1}, P, P^c, A, T) \\ Q &= g_d(Q_{-1}, P_\theta, N, Z) \\ P &= p(P_{-1}, D, I) \\ I &= I_{-1} + Q - D \end{aligned} \qquad (1)$$

Demand D is explained as being dependent on prices P, economic activity A, prices of one or more substitute commodities P^c, and possible technical influences T such as the growth of synthetic substitutes. Other possible influencing factors and the customary stochastic disturbance term are omitted here and elsewhere for simplicity. Accordingly, supply Q would depend on prices as well as natural factors N such as weather, yields, and a possible policy variable Z. A lagged price variable is included since the supply process is normally described using some form of the general class of distribution lag functions. The model is closed using the market clearing identity which equates inventories I with lagged inventories plus supply minus demand. Where the price equation is inverted to represent inventory demand, the identity can be recognized as the equivalent supply of inventories equation.

Among the more important developments related to this methodology as described in Labys (1973) are the incorporation of expectations as reflected in distributed lag structures, other dynamic concepts such as stability, cycle and disequilibrium, use of risk theory in supply response, simulation of models based on experimental design, and optimization experiments using control theory. A well known variant on this form of model is Meadows' (1970) dynamic commodity cycle model. Utilizing an industrial dynamics format, the model's relationships are cast into the form of differential equations with their variables now representing rates of change. It is somewhat restricted to commodities with explicit cycles in production and prices; also emphasized are amplifications and time delays as well as relationships between investment, capacity and output.

Somewhat different from market models, process models deal with supply and demand within an industry rather than across a market; they thus focus on the transformation of commodity inputs into finished products. Whereas market models balance supply and demand to produce an equilibrium price, prices in a process model are normally a function of production and material costs. The emphasis is also different: process models concentrate on the industrial production process, requirements for raw materials and labor and plant capacity. Such a model would begin as follows:

$$\begin{aligned} D &= d(P^P, A) \\ Q^o &= q(L, C, K) = D^o \\ P^o &= P\left(\frac{W \times L}{D^o}, \frac{P \times C}{D^o}\right) \end{aligned} \qquad (2)$$

Whereas the demand for output D^o resembles demand in the market model, supply requires an equation linking the output of a product Q^o to inputs of labour L, raw materials or commodities C, and capital stock K. The identity shows that production is scheduled to equal demand requirements. The final equation explains product prices P^o as a mark-up on prices of commodities P and cost of labour W per unit of output.

Work on these models has mainly been the result of efforts by Adams (1973) and his colleagues. Recent developments include the Adams and Griffin (1972) work on combining the process framework with other model types including macro model estimates of product demand and a linear program which explains the transformation from product output to commodity inputs.

An alternative methodology which integrates the market form of model with a linear programming model is that of spatial equilibrium. Such a model normally consists of three components illustrated as follows: Demand and supply for each market are given first

$$D_j = d(P_j^c, A_j, T_j)$$
$$Q_i = q(P_i, N_i, Z_i) \qquad (3)$$

whereas j indicates the region of consumption and i the region of production. The distribution activities over space or the inter-regional flow of commodities constituting the second component is obtained by decomposing demand and supply into shipments.

$$D_j = \sum_i^n X_{ij} \qquad j = 1, 2, \ldots, n$$
$$Q_i = \sum_j^n X_{ij} \qquad i = 1, 2, \ldots, n \qquad (4)$$

Here X_{ij} denotes the quantities shipped from the ith region to the jth region for n regions. To this must be added the equilibrium component

$$R_{ij}X_{ij} = 0 \qquad \text{all } i,j$$

for which the per unit gain or loss from commodities sold is given by

$$R_{ij} = (P_j - P_i) - TC_{ij} \qquad \text{all } i,j \qquad (5)$$

Since transportation costs TC_{ij} are assumed to be zero within a region, R_{ii} must always equal zero.

While the demand and supply equations imply a structure similar to that of a market model, equilibrium adjustment is more adequately represented through the identification of the profits to be realized from the flow of commodities, i.e. the excess of a price differential between two points minus transportation costs. Profit maximization is assured through the use of linear programming which allows commodities to transfer until demand equals supply in every spatially separated market.

Recent development of this method has stemmed largely from work by Takayama and Judge (1971) who have formulated spatial equilibrium analysis as a quadratic programming problem. There is also the variant on this approach by King and Ho (1965) known as reactive programming. In addition to applying the model to the multicommodity problem, the time dimension has been included in the form of inter-temporal price equilibrium models as well as inter-temporal spatial price equilibrium models. The suitability of the latter for adaptive control analysis as well as progress regarding nonlinear formulations are discussed in Judge and Takayama (1973).

Recursive programming models can be considered a special case of adaptive inter-temporal spatial equilibrium models. They can be described as a sequence of constrained optimization problems in which one or more objective functions, constraint or limitation coefficients of a given problem depend functionally on the optimal primal and/or dual solution vectors of one or more problems earlier in the sequence. However, their practical emphasis on processes of production, investment, and technological change removes them from the main stream of commodity modeling. Day (1973) has largely been responsible for the development of this methodology, having studied problems of substitution during agricultural transition and of technological change in commodity extractive industries. Latest progress relates to constructing multi-sector model as well as to extending the approach to embody general equilibrium.

Also popular is commodity modeling which involves systems analysis in the form of systems models. This normally requires formulating a commodity model so that the major objectives and variables of interest are considered as a complete or functional system. One or more analytical or quantitative methods are combined in a single framework, and a set of decision rules are usually imposed. For example, an econometric model might be integrated with an engineering or biological model as well as include least cost or optimization techniques. Some examples of systems models can be found in the work of Haidacher, et. al. (1975) and in the simulation models of Naylor (1971).

Finally, world trade models can help in studying trade flows and adjustments between commodities. Only of minor interest here, they vary depending on whether the structure of trade or transmissions approach is used in manipulating the included world-trade matrix; system equilibrium is subsequently obtained using an appropriate iterative procedure.

Resource Models

The main interest in commodity resource modeling has been with the mineral commodities; in this section reference will not be made to other types of resource commodities nor to mineral energy commodities. From Table 1, we can obtain an idea of the methodologies most applicable to mineral commodities. The market model shows to have been applied to cobalt and tungsten, the process model to steel, the industrial dynamics model to aluminum and copper, and the recursive programming model to iron and steel together. A full list of model references appears in Labys (1975).

Of these different methodologies, the market model has proven most suitable for further development and application. Here reference is made to the work which has been carried out regarding the non-ferrous metals. Adams (1975) reports on models which have been built successfully for aluminum, chromite, cobalt, copper, lead, manganese, mercury, molybdenum, nickel, platinum, tin, tungsten, and zinc.

Some of the policy application of these models have been reasonably important. For example, the models for aluminum, cobalt, copper, manganese and nickel are being used to study the impact of the production of non-ferrous metal nodules from the ocean floor. As part of the investigation UNCTAD is conducting with regard to future use of the sea bed, the above models have helped to determine the impact of the produced nodules on the price of non-ferrous metals and on the earnings of the producer countries (the latter falling mainly into the group of developing nations). There is also the possibility of further applications pertaining to the analysis of buffer stocks, export quotas and other techniques for stabilizing the prices of these commodities.

However, there are a number of problems to be dealt with before such models can be deemed highly realistic. First of all the models have to be operated together in a multicommodity framework. Substitution and complementarity effects are important on the demand side as well as the supply sides. A model of lead must include linkages with zinc, and a copper model must relate to zinc as well as to lead and tin. Second, some of these substitution effects are better analyzed on the demand side. One possibility for accomplishing this is to consider the demand for these commodities as being interrelated within a linear expenditure system, as recently proposed by Labys and Takayama (1975). Appropriate cross price elasticities could then be determined. Another possibility for studying substitution would be the use of activity analysis describing inter-industry transfers such as that embodied in input-output analysis.

A third problem relates to the supply side. In the short run, the specifications of supply relationships depends primarily on cost factors. But in the long run, supply should depend on factors dealing with resource exhaustion and environmental disturbances. Thus far, commodity models have not dealt with the problems of reserves nor of mechanisms for the recycling of supplies of scrap back into the system. Finally, many of these markets are noncompetitive. Other than work of Burrows (1971), we are just beginning to realize the weaknesses of applying the competitive framework to explain minerals price behaviour.

III ENERGY MODELS

Energy models generally have been constructed from a wide range of methodologies. Since an overall view of energy models and their methodologies has been provided by Limaye (1974), I refer only to particular examples. To begin with, energy models refer to distinct regions. One can think of energy-related commodity models as encompassing a single region, a nation, or the world. Regional models are omitted from discussion here since they belong to the next section.

Of the different methodologies employed in national energy models, the market form of model applied to resource models is also useful here. One of the most elaborate of these models is the Mac Avoy and Pindyck (1972) simulation model of the U.S. natural gas industry. Based on an econometric structure, simulation runs of the model indicate the time path that the endogenous variables would follow given changes in industry and government policies. Process models which concentrate on the transformation between final demands and commodity inputs have been built only for the U.S. petroleum industry. The Adams and Griffin (1972) model of this type has particular advantages for simulation analysis since a linear programming model of petroleum refining technology is placed between sets of relationships describing the demand for gasoline and other refined products on the one hand, and the relationships describing the inputs of crude petroleum on the other. The Day and Tabb (1972) model describing the structure of technological change in the coal industry is the only energy related model of a recursive programming form.

The greatest number of energy models are of a systems type incorporating linear programming or featuring some form of input-output analysis. The Hoffman (1974) linear programming model of the U.S. energy system emphasizes the technological assessment component of planning. Inter-fuel substitutability is taken into account, including both electrical and non-electrical energy. Solution of the L.P. problem yields the optimal demand-supply configuration of the energy system, given the constraints imposed regarding resources, demand, and environmental impacts. In contrast, the Baughman (1974) model of the U.S. energy system studies energy policies based on a combined econometric-mathematical framework. This is closer to a true systems model, in as far as some of the parameters adopted for the model are either assumed or based on technological information. Demand dynamics and capital formation also are emphasized, using coupled equations that connect all primary fuels and secondary sources within a total energy system framework. Input-output analysis also has proven popular. Models have ranged from the Reardon (1972) analysis of U.S. energy consumption to the Hudson and Jorgenson (1974) model which integrates an input-output model with a macro growth model of the U.S. economy.

World energy models obviously because of their complexity have been fewer in number. The Kennedy (1974) world petroleum model embodies the spatial flow methodology with transportation factors now influencing world petroleum distribution. With respect to total energy systems, the Energy Research Unit (1975) model is based on a time-division linear program (TDLP) of the world's oil and gas industry and will soon embody other energy commodities. While neither of these models consider very long run scenarios, the work of Hughes, Mesarovic and Pestel (1974) makes such an attempt. While their model contains only a minimum of behavioural relationships, it is coupled with the group's "Regionalized World Economic Model" which permits long run scenarios to be drawn describing world trade flows, resource exhaustion, pollution, etc.

Where does research in this area of modeling seem to be headed? A certain amount of disinterest has been shown in econometric models because of the instability of their parameters as commodity markets undergo structural change. However, work in this area is progressing, particularly with the use of adaptive control theory as a means of parameter adjustment. In contrast, activity analysis or programming models have not proven as effective in tracing the evolution of a system over time. Many models also have assumed demand and supply to be exogenous, which is often a dubious prescription. With such fragmentation, models have failed typically to include a comprehensive view of the following: demand, supply and price interaction; interfuel competition; and technological and environmental changes. Some of the criticisms raised with respect to resource models would also be applicable here.

IV REGIONAL MODELS

Probably the least viewed context of commodity models is their relation to regional planning. Commodity market models obviously can only serve as a supplement to programming or input-output models of regions. Yet their incorporation in such models should relieve the latter of the assumption that supply and demand are exogenous. Market models can also be embedded in macro planning models as suggested by Labys and Weaver (1973) in their proposed rural development model.

The commodity modeling approach which has had the greatest application in regional planning studies, however, has been that of programming models of the spatial type. The recent publication of Judge and Takayama (1973) has in fact featured a number of model studies of this type. At the one end of the modeling spectrum, we have interregional models for single commodities. These basically explain commodity origin by region of supply and commodity destination by region of demand. Transfers of commodities between regions are then explained by the included network of price differentials and transportation costs, e.g. see Leunis and Vandenborre (1973). Such models also work well in studying interregional trade for the purpose of optimum location of industry, e.g. see Bucholz (1973). At the other end of the spectrum, such models have concentrated on the commodity process within a single region. Models of a multicommodity nature which examine the production potential of a region based on the analysis of farm aggregates, have also been suggested, Weinschenk, _et_._al_. (1973).

But even these models have not been sufficiently comprehensive in dealing with problems of income distribution, employment and resource constraints within a region. It is here that some of the greatest work in coupling commodity models and regional models must be done.

V CONCLUSIONS

This paper is a survey of the commodity modeling approach to resources, energy and regional planning. To those who would call the present methodological taxonomy artificial in terms of commodity models constituting a distinct area, I would respond that such an overview fills an existing need. National models of a macro or interindustry nature generally have proven superior to commodity models in terms of technique and performance. They also have received more interest and attention. Commodity models are just receiving such interest mainly because of the recent international crisis.

REFERENCES

Adams, F.G., 1973, "From Econometric Models of the Nation to Models of Industries and Firms", Wharton Quarterly Business Review.

Adams, F.G., 1975, "Applied Econometric Modeling of Non-Ferrous Metal Markets - The Case of the Ocean Floor Nodules", presented at the RFF Meeting on Minerals Modeling, Washington, D.C.

Adams, F.G. and J.M. Griffin, 1972, "An Econometric Linear Programming Model of the U.S. Petroleum Industry", Journal of the American Statistical Association 67, pp. 542-551.

Baughman, M.L., 1974, "A Model for Energy-Environment Systems Analysis", Energy Modeling (London: IPC Science and Technology Press), pp. 134-149.

Bucholz, H.E., 1975, "Pricing and Allocation Models Applied to Problems of Interregional Trade and Location of Industries", in G. Judge and T. Takayama (eds.), Studies in Economic Planning Over Space and Time (Amsterdam: North Holland Publishing Co.), pp. 298-306.

Burrows, James C., 1971, Tungsten: An Industry Analysis (Lexington, Mass., D.C. Heath and Co.).

Day, R.H., 1973, "Recursive Programming: A Brief Introduction," in G.G. Judge and T. Takayama (eds.), Studies in Economic Planning Over Space and Time (Amsterdam: North Holland), pp. 329-344.

Day, R.H. and W.K. Tabb, 1972, "A Dynamic Microeconomic Model of the U.S. Coal Mining Industry", SSRI Research Paper, University of Wisconsin, Madison.

Energy Research Unit, 1974, "World Energy Model: Description and Results", in Energy Modeling (London: IPC Science and Technology Press, Ltd.).

Haidacher, R.C., Kite, R.C. and Matthews, J.L., 1975, "Application of a Planning-Decision Model for Surplus Commodity Removal Programs", in W.C. Labys (ed.) Quantitative Models of Community Markets (Cambridge: Ballinger Publishing Co.), pp. 265-290.

Hoffman, K.C., 1972, "The United States Energy System - A Unified Planning Framework", Unpublished Ph.D. dissertation, Polytechnic Institute of Brooklyn,

Hudson, E.A. and Jorgenson, D.W., 1974, "U.S. Energy Policy and Economic Growth, 1975-2000", Bell Journal of Economics and Management Science 5, pp. 461-514.

Hughes, B., Mesarovic, M., and Pestel, E., 1974, "Energy Models: Resources, Demand and Supply", Multilevel Regionalized World Modeling Project, Case Western Reserve University, Cleveland.

Judge, G.G. and T. Takayama (eds.), 1973, Studies in Economic Planning Over Space and Time (Amsterdam: North Holland).

Kennedy, M., 1974, "An Economic Model of the World Oil Market", Bell Journal of Economics and Management Science 5, pp. 540-577.

King, R.A. and Foo-Shiung Ho, 1965, "Reactive Programming: A Market Simulating Spatial Equilibrium Algorithm", Economics Special Report, Department of Economics, North Carolina State University at Raleigh.

Labys, W.D., 1973, Dynamic Commodity Models: Specification, Estimation and Simulation (Lexington: Heath Lexington Books).

Labys, W.D. (ed.), 1975, Quantitative Models of Commodity Markets (Cambridge: Ballinger Publishing Co.).

Labys, W.C. and Takayama, 1975, "Measuring Multicommodity Substitution Patterns with a Demand System Approach", Mimeographed, Graduate Institute of International Studies, Geneva.

Labys, W.C. and T.F. Weaver, 1973, "Towards a Commodity Oriented Development Model", SEADAG Paper, Seminar on Directions in Rural Development Planning, New York.

Leunis, J.V. and R.J. Vandenborre, 1973, "An Interregional Analysis of the U.S. Soybean Industry", in G. Judge and T. Takayama (eds.), Studies in Economic Planning Over Space and Time (Amsterdam: North Holland Publishing Co.), pp. 274-297.

Limaye, D.R., 1974, Energy Policy Evaluation (Lexington: Heath Lexington Books).

Mac Avoy, P.W. and R.S. Pindyck, 1972, "Alternative Regulatory Policies for Dealing with the Natural Gas Shortage", Bell Journal of Economics and Management Science 3, pp. 454-498.

Meadows, R.H., 1971, Computer Simulation Experiments with Models of Economic Systems (New York: John Wiley & Sons).

Reardon, W.A., 1972, "An Input/Output Analysis of Energy Use Changes from 1947 to 1958 and 1958 to 1963", (BATTELLE, Pacific Northwest Laboratories).

Takayama, T. and G.G. Judge, 1971, Spatial and Temporal Price and Equilibrium Models (Amsterdam: North Holland Publishing Co.).

Weinschenk, G., Heinrichsmeyer and C.H. Hanf, 1975, "Experiences with Multicommodity Models in Regional Analysis", in G. Judge and T. Takayama (eds.), Studies in Economic Planning Over Space and Time (Amsterdam: North Holland Publishing Co.), pp. 307-328.

TABLE 1

COMMODITY MODELING METHODOLOGIES AND THE MODELING PROCESS

Modeling Process / Methodologies	What do the Methodologies describe?	What quantitative method is used?	What economic behavior is specified	Examples of Commodity Applications
Market Model	Demand, supply, inventories interact to produce an equilibrium price in competitive or non-competitive markets	Dynamic micro econometric system composed of difference or differential equations	Interaction between decision makers in reaching market equilibrium based on demand, supply, inventories, prices, trade, etc.	Cobalt(42) Energy(30) Lauric Oils(51) Soybeans (261) Sugar(200) Tungsten (298)
Process Model	Demand and production within an industry, focussing on transformation from product demand to input requirements	Dynamic micro econometric difference equation system suitable for integrating linear programming on production side	Interaction between decision makers in industries, markets, national economies based on demand, inventories, production, investment, capacity utilization, commodity inputs, prices, etc.	Petroleum (227) Steel(269)
Dynamic Commodity Cycle Model-Industrial Dynamics Model	Demand, supply, inventories interact to produce an equilibrium price emphasizing role of amplifications and feedback delays	Dynamic micro econometric differential equation system which features lagged feedback relations and variables in rates of change	Interaction between decision makers in adjusting rate of production to maintain a desired level of inventory in relationship to rate of consumption	Aluminum (2) Broilers (33) Cattle(166) Copper(2) Hogs(142) Orange Juice(218)
World Trade Model	Imports and exports balance between regions given adjustments in income. Transmissions versus structure of trade approach depends on use of trade matrix	Macro/micro econometric equation system with equilibrium obtained in a simulation framework through an iterative procedure	Interaction between decision makers in markets and national econmies in reaching equilibrium based on adjustments in imports, exports, prices and national income	No disaggregated models at present

Table 1 (continued)

Modeling Process / Methodologies	What do the Methodologies describe?	What quantitative method is used?	What economic behavior is specified	Examples of Commodity Application
Spatial Equilibrium Model	Spatial flows of demand and supply and equilibrium conditions assigned optimally in equilibrium depending on configuration of transportation network	Activity analysis of a spatial and/or temporal form. Degree of complexity depends on endogeneity and method of incorporation of demand and supply functions	Interaction between decision makers in allocating shipments (exports) and consumption (imports) optimized through maximizing sectoral revenues or minimizing sectoral costs	Bananas(12) Broilers(32) Livestock (155) Oranges(215) Palm Oil(105) Wheat(317)
Recursive Programming Model	Production conditions and input revenue determined through primal/dual of linear program. Recursivity introduced through feedback component which includes profit, capital and demand	Activity analysis involving a sequence of constrained maximization problems in which objective function limitation coefficients depend on optimal primal/dual solutions attained earlier in the sequence	Interaction between decision makers in reaching market equilibrium involves adaptive intertemporal processes related to production investment and technological change	Coal (39) Iron,steel (272) Wheat,corn, soybeans (309)
Systems Model	Demand, supply and other major variables and objectives considered as a complete system rather than a single market	Dynamic micro econometric equation system which when formed into a simulation framework is coupled with activity analysis and/or decision rules	Interaction between decision makers belonging to the system environment based on performance variables such as revenues, costs as well as market variables such as demand, supply, prices, etc.	Beef (21) Energy(85) Fish (124) Livestock (161) Multicommodity(205) Rice (243)

DETERMINATION OF SOCIAL COSTS OF ENVIRONMENTAL DAMAGE

A.P. Mastenbroek and P. Nijkamp

I INTRODUCTION

From the end of the sixties onwards, regional science has increasingly focussed its attention on environmental problems. Both the extent and the intensity of these environmental problems in western industrialized areas have urged regional scientists to consider environmental quality analysis as an integral part of their profession. One is increasingly becoming aware of the fact that the spatial pattern of entrepreneurial and residential locations, prevailing technology and environmental quality possess an intricate interwovenness. In certain areas and under certain prevailing technologies fresh air, clean water and a pleasant environment tend to become scarce goods. Since economics is particularly oriented towards the study of scarcity, environmental quality analysis can obviously benefit from the use of economic tools.

A profound economic analysis of environmental problems provides more insight into the economic causes and consequences of pollution phenomena. The rapid growth of national production, the use of synthetic raw materials, the increased use of cars and the spatial concentrations of industries and population have shaped the conditions under which a large-scale pollution could arise. Furthermore, during the post-war decades the strong priorities of western societies for material welfare neglected the consequences of the economic and technological growth for man's physical environment.

In economic terms, environmental disruption stems from so-called _externalities_, which can be considered as side-effects of human activities which affect the allocation of scarce resources, but which are not fully reflected in the price mechanism. The classical mechanism presupposed a rational confrontation of demand and supply, such that a market equilibrium reflects all sacrifices made by market participants. In case of pollution, however, environmental disruption is not included as a cost element in entrepreneurial and consumer decisions, so that the actual costs of many commodities are underestimated and a biased market equilibrium is attained.

In order to avoid such a bias, the aforementioned externalities should be _internalized_. This implies that the costs of environmental damage have to be included in economic decisions, such that social preferences for maintenance of environmental quality are included in the production and consumption decisions.

This paper concentrates on the latter problem. After a survey of a set of methods for evaluating environmental deteriorations in economic terms, a new method will be developed to measure the 'optimal' social costs of environmental damage from air pollution. This method, based on a so-called implicit optimal approach, attempts to introduce the implicit social costs of pollution standards into the usual environmental programming models. These optimal damage costs of pollution are then calculated by means of iterative approximation procedures. The method is illustrated by means of some numerical applications, while finally some further extensions are proposed.

II STRATEGIES FOR EVALUATING ENVIRONMENTAL COMMODITIES

In the first paragraph some introductory remarks were presented concerning environmental deterioration. Now the question arises: how to evaluate man's physical environment as well as the change in it?

The evaluation of environmental phenomena rests essentially on rank orders of preferences for environmental commodities and other commodities. An environmental commodity can be conceived of as any element associated with man's physical environment which is able to influence man's relative welfare position, although it is not necessarily priced at a market.

A first method to arrive at a numerical evaluation of environmental commodities starts from general welfare economics. This method, called the implicit method (Nijkamp and Somermeijer [1971]), is based on a revealed preference hypothesis and it aims at gauging implicitly the relative evluations of sets of commodities. On the basis of actual choice behaviour the implicit values of preference parameters may be inferred, assuming that actual decisions reflect the decision-maker's implicit relative preferences.

The aforementioned implicit method may be illustrated for environmental problems by assuming that national (regional) income can be subdivided into two categories, viz. 'environmental' expenditures (z) and other expenditures (y). 'Environmental' expenditures are spent of environmental goods; they are among others composed of anti-pollution investments, investments in parks and in recreation facilities, and consumption expenditures to environmental commodities (recreation, e.g.). Next, one may assume that the total national (regional) resources r can be used to generate y and z in alternative combinations, which can be represented by means of a product transformation curve (or production possibilities curve). This curve is the locus of the maximum combinations of z and y that can be produced with given resources and technology. The slope of this curve represents the social marginal opportunity cost.

If the transformation curve is assumed to be linear, it can be represented as:

$$\eta y + \theta z = r \tag{2.1}$$

Assuming a welfare criterion $\omega(y, z)$, the following optimizing program is obtained:

$$\left.\begin{array}{l} \max \ \omega(y, z) \\ \text{subject to} \\ \eta y + \theta z = r \end{array}\right\} \tag{2.2}$$

The first-order conditions for a maximum provide (the absolute value of) the social marginal opportunity cost of environmental goods with respect to other goods:

$$\frac{\nabla_y}{\nabla_z} = \frac{\eta}{\theta} \tag{2.3}$$

where ∇_y and ∇_z are the gradient (first-order derivative) of $\omega(y, z)$ with respect to y and z, respectively.
Relationship (2.3.) can be rewritten as a ratio of welfare elasticities:

$$E_{yz} = \frac{\nabla_y \frac{y}{\omega}}{\nabla_z \frac{z}{\omega}} = \frac{\nabla_y}{\nabla_z} \frac{y}{z} = \frac{\eta}{\theta} \frac{y}{z} \tag{2.4}$$

or using a proportionality symbol $\propto$:

$$E_{yz} \propto \frac{y}{z} \tag{2.5}$$

Assuming that the actual values of y and z can implicitly be conceived of as approximations for the value of environmental goods and other goods, an *ex post* analysis can be applied to estimate the evolution of the relative social evaluation of environmental commodities. An application of this method is contained among others in Nijkamp and Paelinck [1973].

The previous collective approach to the evaluation of environmental commodities is carried out at an aggregated level. It provides insight into welfare implications and prevailing trends of collective evaluations of environmental commodities, but it can hardly be utilized at a disaggregated level to gauge the net social benefits of individual or local environmental commodities.

The same holds true for another method, viz. an actual or fictitious interview procedure to determine the marginal rates of substitution and hence the net social benefits of alternative combinations of environmental goods and other goods. Therefore, the majority of operational environmental analyses is not based on the previous collective welfare approach.

To obtain monetary estimates, decision criteria like the maximization of (national or regional) income of all sectors or the minimization of environmental damage are useful to judge the state and development of the economy. Such more manageable decision criteria will be introduced later on in this paper to measure certain cost elements of air pollution.

It is clear, however, that obtaining a more disaggregated view of environmental phenomena requires a more detailed insight into the various effects of environmental deterioration. A manageable way of determining these effects is the use of a so-called impact analysis. An impact analysis provides a detailed picture of the consequences of economic and technological decisions for ecological, spatial and economic structures and processes. For example, the construction of a new highway leads to a rise in demand for labour, an increased space occupation, a rise in accessibility, and a destruction of natural areas. A calculation of all separate effects requires a detailed representation of the structure of the phenomena at hand.

In applying such an impact analysis it is useful to distinguish two phases, viz. the purely technical-physical effects of a certain phenomenon (air pollution, e.g.) and the economic evaluation of these effects. For example, the pollution vector $\underline{p}$ associated with a certain production vector $\underline{q}$ can be represented by means of the following impact function $\underline{f}$:

$$\underline{p} = \underline{f}(\underline{q}) \tag{2.6}$$

while next the spatial concentration $\underline{c}$ of pollution can be represented by means of a dispersion function $\underline{h}$:

$$\underline{c} = \underline{h}(\underline{p}) \tag{2.7}$$

The vector of physical environmental effects $\underline{d}$ accrueing from the pollution level $\underline{p}$ can be represented by means of the effect function $\underline{g}$, which represents the physical effect to each object affected as a function of the intensity of air pollution:

$$\underline{d} = \underline{g}(\underline{c}) \tag{2.8}$$

Relationship (2.8.) can be written in general notation as:

$$\underline{d} = \underline{g}[\underline{h}\{\underline{f}(\underline{q})\}], \tag{2.9}$$

which is essentially a stimulus-effect representation of a set of consequences to various environmental objects accruing from a certain production structure.

The previous relationships (2.6.) - (2.9.) are of a technical nature: they have nothing to do with economic evaluations.

Now the question arises: how to gauge the environmental impacts of a certain production (and consumption) structure in economic terms? In other words: which monetary weights $\underline{\mu}$ have to be assigned to the effect vector $\underline{d}$? The answer to the latter question would imply that the total economic value m of environmental impacts (measured as net social benefits) from production (and consumption) decisions would be equal to:

$$\begin{aligned} m &= \underline{\mu}' \underline{d} \\ &= \underline{\mu}' \underline{g}[\underline{h}\{\underline{f}(\underline{q})\}] \end{aligned} \qquad (2.10)$$

where m is the aggregated value of physical effects to environmental goods (Ridker [1966]).
The latter equation implies that all different physical consequences to environmental commodities (with different dimensions) are brought into relation with the measuring rod of money (Pigou [1920]).

Therefore, in principle relationship (2.10.) could be used to arrive at estimates of the total cost of environmental deterioration (or, in general, of the total net social benefits of economic impacts on environmental commodities). Such a monetary evaluation of environmental effects measures in fact the loss of benefits in economic functions of environmental commodities.

Applications and discussions of the aforementioned approach with a variety of adaptations and modifications are contained among others in Bain [1973], Freeman et al. [1973], Hueting [1974], Jansen et al. [1974], Mishan [1971], Opschoor [1974], Ridker [1967] and Wolozin [1966].

Frequently, this approach served to evaluate the monetary damage of environmental pollution.

The damage costs of environmental pollution may relate to various objects affected. Examples are:
- damage to health (Klarman [1965] and Lave and Seskin [1971]).
- damage to residential properties (Crocker [1972], Jaksch and Stoevener [1970], Ridker and Henning [1967]).
- annoyance due to noise of aircrafts or traffic (Cantilli [1974], Plowden [1970] and Stradford [1974]).
- damage to agricultural production (Barrett and Waddell [1970]).
- damage to industrial installations due to corrosion (Barrett and Waddell [1970]).
- damage to natural areas from infrastructural investments (Klaassen and Verster [1974] and Nijkamp [1974]).

The question arises, however, whether the foregoing strategy for measuring the net social costs (or benefits) of environmental pollution leads to reliable and useful outcomes. Therefore, the damage cost vector $\underline{\mu}$ should be inspected more accurately. For each object affected the corresponding unit damage cost can be gauged in two complementary ways.

First, certain cost elements can be based on direct damage to private or collective properties (for example, damage to agriculture from polluted air and water, and destruction of natural areas due to the construction of a new highway). Instead of direct damage costs one could also use the amounts of money to be paid in order to compensate economic subjects for their decline in welfare (for a discussion of the compensation principle, see Coase [1960], Goudzwaard [1970], Little [1950], Mishan [1967], and Pigon [1920]).

This first cost category can be represented in formal terms by means of a damage function d_A with environmental quality v as an argument (cf. Dales [1968], Freeman et al. [1973], Mastenbroek and Nijkamp [1974], and Ridker [197]]):

$$d_A = f(v) \tag{2.11}$$

It is evident that d_A is an inverse function of v; the following assumptions are generally made about the shape of the function (Figure 1.):

$$\left.\begin{aligned} \frac{\partial d_A}{\partial v} &< 0 \\ \frac{\partial^2 d_A}{\partial v^2} &\geqq 0 \end{aligned}\right\} \tag{2.12}$$

Second, other cost elements of environmental damage can be based on adjustments to reduce the direct impact of pollution (for example, by migrating to a region with a cleaner environment, or by implementing abatement investments like waterpurifying installations). The curve of pollution prevention costs (abatement costs, control costs) d_B can be represented formally as:

$$d_B = f(v), \tag{2.13}$$

where d_B and v are obviously positively correlated. The following assumptions can be made about function (2.13.) (cf. figure 3.1.):

$$\left.\begin{aligned} \frac{\partial d_B}{\partial v} &> 0 \\ \frac{\partial^2 d_B}{\partial v^2} &\geqq 0 \end{aligned}\right\} \tag{2.14}$$

An illustration of the slopes of curves (2.11.) and (2.13.) is found in figure 1.1.

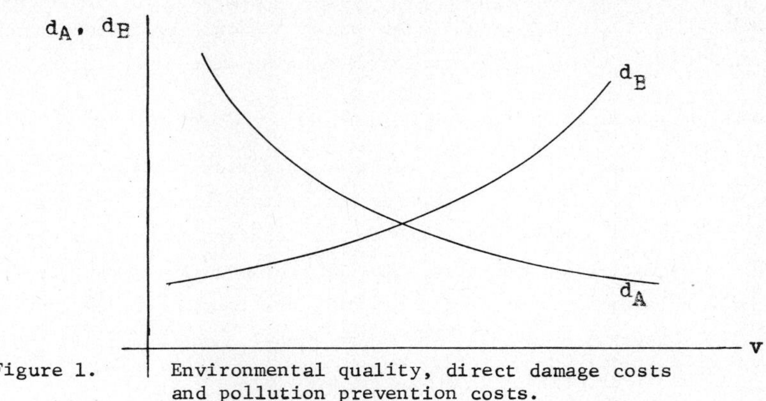

Figure 1. Environmental quality, direct damage costs and pollution prevention costs.

It is clear that an aggregation of all individual cost figures for all economic activities and all objects affected is fraught with difficulties and may involve many disturbances and deviations:

- An aggregation of individual cost figures is only meaningful if a bulk of information and data on pollution and environmental effects is available (including meteorological data).
- Environmental quality is a multidimensional phenomenon (both with regard to objects and to disturbances), so that one cannot expect a priori to find a unidimensional proxy for environmental quality.
- A constant cost per unit damage (cf. (2.10.)) is hardly acceptable in view of Figure 2.1., particularly if there is no market price that can serve as a useful approximation.
- The determination of total damage costs is based on a static approach in which future effects of environmental deterioration (for example, accumulation of pollutants) are neglected.
- The spatial dispersion of economic activities and their effects on the pollution pattern are assumed to be given, so that neither displacements of activities (cf. the Dutch selective investment act) nor distributional aspects of environmental deterioration are considered.

In spite of the foregoing drawbacks of the various strategies for measuring the value of environmental commodities, particularly at a local, disaggregated level, these methods may be meaningful to shed some light on the social sacrifices necessary to achieve or to maintain a certain production and consumption pattern. A serious problem, however, in employing the previous methods is the question how to use the various cost calculations in environmental administration. In general, there is no feedback between the cost figures and pollution control. In a next paragraph a proposal will be made to use damage costs as shadow elements to arrive at an optimal environmental administration. By means of this method many drawbacks of the individual cost approach can be attacked.

III A STRATEGY FOR OPTIMAL POLLUTION CONTROL

In this paragraph attention will be paid to the problem of optimal collective decision making in the field of pollution control. The assumption will be made that man's physical environment can be conceived of as a common property good. Environmental deterioration implies that the present and future consumption of environmental goods (cf. section II.) is threatened or affected by the technology employed and the prevailing consumption pattern. Without public intervention (pollution charges, e.g.) the quantity and quality of environmental goods would undergo an undesirable decline. In this respect, the notion of a merit good (cf. Musgrave [1969]), might be relevant, which indicates that, even in case of a market equilibrium, a certain good would be insufficiently available in the opinion of public decision makers. Given this assumption, it seems to be reasonable for public authorities to specify minimum threshold levels for the availability or quality of environmental goods (effluent standards e.g.). These threshold levels need not only result from human health conditions, but also from more general (non-antropocentric) stability conditions of fragile ecosystems. It should be noted, that these threshold levels rest essentially on a political decision.

The purpose of our analysis is to measure the social costs of different levels and types of air-pollution controls. By experimenting with alternative levels of controls, the effects on regional welfare will be calculated as well as the corresponding optimal production structure. This optimal production structure is based on the social costs of air-pollution controls by means of a shadow price approach. First, an appropriate environmental model will be set out in more deatil.

For a particular region, the appropriate collective decision-criterion is assumed to be the maximization of regional economic efficiency (i.e. regional value added minus the opportunity costs of environmental threshold levels). Therefore, the formal specification of the decision criterion at hand is:

$$\max \omega = \underline{v}'\underline{q} - \underline{\lambda}'\underline{p} \qquad (3.1)$$

where $\underline{v}$, $\underline{q}$, $\underline{\lambda}$ and $\underline{p}$ represent vectors of (known) regional value-added coefficients per sector, (unknown) regional production levels per sector, (unknown) unit costs of a set of pollutants, and (unknown) pollution levels, respectively. Next, the following effluent standards v $\underline{p}^*$ for the vector $\underline{p}$ of pollutants are assumed to be relevant:

$$\underline{p} \leq \underline{p}^*$$

Obviously, the use of effluent standards provides no incentive to reduce residuals discharges beyond the permissible level, but the use of effluent standards appears to be more effective than a system of charges which is very hard to specify (cf. Kneese [1965] and Victor [1972]). The upper limits on the successive emissions (or concentrations) of pollutants are frequently specified on the basis of medical, biological and ecological information. The basic difference of our approach with respect to alternative models in this field (cf. Burton and Sanjour [1970] and Kohn [1970]) is, that the social costs of pollution are directly subtracted from regional income. This less optimistic view affects obviously the optimal regional production structure based on a single value-added criterion (cf. also Russell and Spofford [1972]).

A serious problem, however, is the fact that the pollution costs are in general unknown, since many cost data are not available. Therefore, the attention will be focused here on the shadow prices of environmental deterioration. These shadow prices reflect the social costs inherent to the elimination of pollution towards a permissible threshold level of emission. The total shadow costs of environmental pollution can be considered as an approximate monetary social evaluation. A formal approach to shadow prices is contained in a next paragraph. By using this shadow price approach the social costs of environmental externatlities are introduced into collective economic decision making.

It is obvious that a further analysis requires the use of a regional industry model that relates inputs and outputs of the various production processes and consumption activities, as well as the amounts of pollutants generated by each activity. Assuming a linear industry model for production inputs and outputs, the following input-output framework may be specified:

$$\underline{q} = A\underline{q} + \underline{f} \qquad (3.3)$$

where A and $\underline{f}$ are an input-output matrix and a final demand vector, respectively. Furthermore, the residuals generated by a certain production structure can be calculated as:

$$\underline{p} = B\underline{q} \, , \qquad (3.4)$$

where B is the matrix of pollution input-output coefficients (Leontief [1970] and Isard [1972]).

Next, environmental dispersion models (of a Gaussian type, e.g.) can be used to transform the residuals discharges into ambient concentrations $\underline{c}$ at various points throughout the environment (Pasquill [1962]):

$$\underline{c} = \underline{h}(\underline{p}) \, , \qquad (3.5)$$

which corresponds to (2.7.). It is obvious, that the threshold levels for a maximum permissible pollution could also be imposed on the ambient concentrations $\underline{c}$

instead of on the residuals discharges $\underline{p}$. This would require a corresponding adaptation of the objective function (3.1.), although the essential structure would remain identical.

The foregoing model can be extended in several ways (Mastenbroek en Nijkamp [1974]). First, residuals treatment processes can be introduced, so that the degree of pollution is co-determined by the amount of abatement investments (filters, water purifying installations, etc.). This would imply that the input-output model (3.3.) can be rewritten as:

$$\underline{q} = A\underline{q} + \underline{i}^A + \underline{i}^P + \underline{d} + \underline{x} \, , \qquad (3.6)$$

where the final demand vector $\underline{f}$ has been divided into 4 components $\underline{i}^A$, $\underline{i}^P$, $\underline{d}$ and $\underline{x}$, which represents the amount of sectoral abatement investments, the amount of remaining ('productive') investments (directly earmarked for production), the volume of sectoral consumption, and net exports, respectively.

By assuming that the residuals discharges are an inverse function of the amount of abatement investments, relationships (3.4.) can be written as (cf. ENVIRONMENT [1974]):

$$\underline{p} = B\underline{q} - \underline{f}(\underline{i}^A) \qquad (3.7)$$

where $\underline{f}$ represents the residuals treatment processes associated with anti-pollution investment $\underline{i}^A$.

Next, an appropriate extension of the previous model is obtained by introducint a (simplified) investment equation, which links $\underline{i}^P$ to $\underline{q}$:

$$\underline{\Delta q} = \underline{q}_t - \underline{q}_{t-1} = \hat{K}^{-1} \underline{i}_t^P \qquad (3.8)$$

where $\hat{K}$ is a diagonal matrix with marginal capital-output ratios as diagnoal elements, and where the lower index t represents the relevant time period; $\underline{q}_{t-1}$ is supposed to be known at the beginning of the decision strategy. Finally, in order to avoid an excessive level of realized investments (particularly in the abatement sector), an upper limit $\underline{i}^*$ on the total investment budget is imposed:

$$\underline{i}^A + \underline{i}^P \leq \underline{i}^* \qquad (3.9)$$

It is obvious, that the previous model can still be extended in many ways, for instance by dealing with employment and land use constraints. For example, an upper limit 1^* on total labour supply can be specified as:

$$\underline{1}'\underline{q} \leq 1^* \, , \qquad (3.10)$$

where $\underline{1}$ is a vector of (constant) employment coefficients for each individual sector. In a similar way, a maximum land use constraint s^* can be included as:

$$\underline{s}'\underline{q} \leq s^* \qquad (3.11)$$

where $\underline{s}$ represents a vector of average land use coefficients per sector. The previous model is essentially controlled by some counteracting forces. First, the positive term $\underline{v}'\underline{q}$ in the objective function (3.1.) will induce a situation where $\underline{q}$ will tend to be high. Second, the negative term $-\underline{\lambda}'\underline{p}$ (supposed $\underline{\lambda}$ is known) will induce a situation where $\underline{p}$ is at a minimum. An equilibrium may be expected at an intermediate position, where $\underline{p}$ does not exceed $\underline{p}^*$ and where regional income is as high as possible. This implies a tendency towards a situation where many effluent standards are active.

The foregoing model is essentially a specific programming model. Should all relationships be specified linearly, then this model is a particular linear programming (L.P.) model. Now the problem arises: how to solve this problem, given the fact that the unknown vector $\underline{\lambda}$ in (3.1.) plays a crucial role as environmental pollution costs? If one assumes that $\underline{\lambda}$ is an unknown cost vector, a very serious difficulty arises. Then the foregoing programming model should try to identify both the optimal production, consumption and pollution levels, and the shadow price vector of the corresponding optimal pollution level.

The foregoing problem can in principle be solved by using the effluent standards specified in (3.2.). For the moment the assumption is made that there exists only one system of effluent standards, so that trade-offs between different systems (for example, on the basis of cost-minimizing combinations) need not to be considered. A system of alternative effluent standards will be considered later on.

The effluent standards will now be used as key elements to determine the unknown cost vector $\underline{\lambda}$. This vector can be conceived of as the implicit social evolution of a certain pollution level $\underline{p}$, or more precisely as the marginal social costs (i.e., the decline in regional income) of reducing pollution with one additional unit of a certain pollutant (as far as the pollution level exceeds the effluent standards). In general, one or more of the effluent standards in the foregoing model will be active due to the income maximizing objective function. Given the latter assumption, $\underline{\lambda}$ can be conceived of as the shadow price (Lagrange multiplier) associated with a pollution level the size of the effluent standard. Now the problem arises as to how to find an optimal shadow price $\underline{\lambda}$ which reflects the implicit social evaluation of a certain environmental quality reflected by the effluent standard $\underline{p}^*$.

The value of the unknown shadow price $\underline{\lambda}$ associated with the environmental quality constraints can be determined by employing a recursive procedure. A first approximation of $\underline{\lambda}$ can be found by substituting for $\underline{\lambda}$ the vector of Lagrange multipliers associated with the effluent standards (3.2.). This Lagrange multiplier can in principle be calculated from any optimization program, and even very easily from a L.P. program. By substituting the vector of Lagrange multipliers into (3.1.), a new optimization model is obtained, which can be solved in a similar way. On the basis of this optimization model a new set of Lagrange multipliers can be calculated, which can be used as a second-round approximation of $\underline{\lambda}$, etc. The previous iterative scheme is continued, until a convergence up to a desired degree of accuracy is attained. This recursive procedure leads to an optimal set of equilibrium values for $\underline{q}$, $\underline{p}$ and $\underline{\lambda}$ (for a formal presentation, see IV).

The previous iterative shadow price approach does not measure the value of environmental commodities, but only the implicit social values of certain environmental quality standards. Therefore, the shadow price approach can be interpreted by means of Lagrangian (or Kuhn-Tucker) theory. It is based on the fact that a shadow price represents the marginal change in an objective function due to a shift in a corresponding (active) constraint.

It should be noted that the optimal implicit social costs of establishing environmental quality standards are based on active effluent standards. Should certain pollution standards be inactive, then the shadow price of the corresponding pollutant is equal to zero, since the pollutant concerned does not exert any influence on the variables within the economic system in question. The latter situation is closely related to the traditional economic way of thinking, that free goods (fresh air, e.g.) do not possess a market value. Therefore, it is necessary that the effluent standards are established at rather low levels, so that even in case of minor environmental damages the classical welfare indicator, viz. per capita income, is corrected for pollution costs and the environmental policy is used as a tool to change the production structure. Then the shadow prices of effluent standards receive a meaningful interpretation, viz. the normative opportunity costs of a cleaner environment.

Finally, the shadow price of an effluent standard will, in general, be different for each separate region. This implies that the social evaluation of environmental quality standards is determined by the regional economic structure, the regional welfare function and the effluent standards.

Finally, some attention has to be paid to the level of the effluent standards. As exposed before, the determination of these standards is ultimately a matter of political responsibility, although based on scientific inquiry from different disciplines.

A more adequate insight into the effect of different effluent standards on the net regional benefit is obtained by repeating the aforementioned recursive procedure for a set of different effluent standards. This implies that for each level separately a corresponding shadow price can be calculated, so that ultimately a cost function for imposing various effluent standards can be derived. This cost function will, in general, be a kinked curve, since the use of L.P. models implies that only corner points are relevant solution points. By varying the constraints of an L.P. program the optimal solution may switch to an other corner point. A numerical illustration of such a parametric programme is contained in section V, whereas in par. 6 an application of this method is presented.

IV FORMAL PROPERTIES OF THE ENVIRONMENTAL QUALITY MODEL

In the foregoing paragraph a method was developed for determining 'optimal' social costs of emission standards. In this paragraph the latter method will be considered from a more formal point of view. The essential structure of this decision problem was formed by the maximization of a net economic surplus, given a set of constraints (equalities as well as inequalities) on the economic structure in question. The purpose of the model was to calculate the costs (i.e., loss in income) due to the presence of effluent standards. Therefore, the formal general presentation of the decision problem specified in (3.1.) - (3.11.) is:

$$\left.\begin{array}{l} \max \omega = \underline{v}'\underline{q} - \lambda'\underline{p} \\ \text{s.t.} \\ C\,\underline{x} = \underline{c} \\ D\,\underline{x} \leq \underline{d} \\ \underline{x} \geq \underline{0} \end{array}\right\} \qquad (4.1)$$

where $\underline{x}$ is a general vector which encompasses all variables of the decision problem in question (including $\underline{q}$ and $\underline{p}$), C a matrix of coefficients corresponding to the equality constraints and $\underline{c}$ a vector of corresponding constant terms. In a similar way, D and $\underline{d}$ represent the coefficient matrix and the constant terms of the inequality conditions. Obviously, $\underline{d}$ encompasses also the effluent standards for all pollutants (cf. also (3.2.)).

If necessary, the previous linear system can be tranformed into a nonlinear specification (Nijkamp and Paelinck [1973]), into a dynamic specification (Nijkamp [1974]) or into a stochastic specification (particularly as far as the diffusion of the pollutants is concerned).

As set out in the previous paragraph, in the first step of the iteration procedure $\underline{\lambda}$ is set equal to zero. Therefore, the decision problem at hand is now a normal L.P. program. It is a well-known property of L.P. models that the solution algorithm attempts to identify such a set of <u>active</u>, independent constraints that the objective function is at a maximum. The resulting system of active constraints constitutes a square system of linearly independent equations (Nijkamp and Paelinck

[1975]). Since the effluent standards impose a limit on a further extension of the production, it is clear that many of these effluent standards will be active at the optimum.

The previous property of the solution algorithm implies that at the optimum only the following set of equality conditions is relevant:

$$\begin{bmatrix} C \\ \hline D^* \end{bmatrix} \underline{x} = \begin{bmatrix} \underline{c} \\ \hline \underline{d}^* \end{bmatrix} \tag{4.2}$$

where D^* and $\underline{d}^*$ are only a subset of the parameters of the original inequality constraints (including the non-negativity conditions), viz. those parameters which correspond to active constraints. The coefficient matrix is now a square matrix. The remaining inequality constraints are satisfied, but they do not play an active role.

Given side-condition (4.2.), the dual L.P. model can be written as:

$$\min \quad = \underline{\mu}'\underline{c} + \underline{\nu}'\underline{d}^*$$

s.t.

$$\left[C' \;\vdots\; D^{*\prime} \right] \begin{bmatrix} \underline{\mu} \\ \\ \underline{\nu} \end{bmatrix} = \begin{bmatrix} \underline{e} \end{bmatrix} \tag{4.3}$$

where $\underline{\mu}$ and $\underline{\nu}$ are the (non-zero) shadow prices associated with the constraint vector from (4.2.), and where $\underline{e}$ is a general vector of parameters resulting from the primal objective function (i.e. $\underline{e}$ includes $\underline{v}$).

It is evident, that $\underline{\mu}$ and $\underline{\nu}$ can directly be calculated from (4.3.), so that the shadow variables of all active primal constraints are known. This implies that the vector $\underline{\lambda}$ from (4.1.) is also known, since the non-zero elements of this vector are a subset of $\underline{\nu}$. It should be noted that $\underline{\lambda}$ has only non-zero elements for corresponding active constraints. The subvector of $\underline{\lambda}$ which includes non-zero elements will be denoted by $\underline{\lambda}^*$.

The second iteration of the previous L.P. model is carried out after substitution of the initial value of $\underline{\lambda}$, denoted by $\underline{\lambda}_1$, into (4.1.). This requires the solution of a new L.P. program, which will give different results. A new value of $\underline{\lambda}$, denoted by $\underline{\lambda}_2$, can again be calculated from (4.3.). This new value is again substituted into (4.1.), etc.

It should be noted that from the second variation onwards the parameter vector $\underline{e}$ from (4.3.) contain also the value of $-\underline{\lambda}$ from the previous iteration.

Should there be no switch among active constraints from a certain step onwards, then the equilibrium value of $\underline{\lambda}$ can directly be calculated from the constraint set of (4.3.). In the latter case, both the left-hand side and the right-hand side of this linear system contain $\underline{\lambda}^*$, so that the optimal value of $\underline{\lambda}^*$ can easily be found by solving the linear system in question. This is easily seen by including $\underline{\lambda}$ into the constraint set of (4.3.) as follows:

$$\left[\begin{array}{c|c} C' & D^{*\prime} \end{array} \right] \begin{bmatrix} \underline{\mu} \\ \hline \underline{\nu}^* \\ \hline \underline{\lambda}^* \end{bmatrix} = \begin{bmatrix} \underline{e}^* \\ \hline -\underline{\lambda}^* \end{bmatrix} \tag{4.4}$$

where the new vectors $\underline{\nu}$ and $\underline{e}^*$ result from the following definitions:

$$\begin{bmatrix} \underline{\nu}^* \\ \hline \underline{\lambda}^* \end{bmatrix} = \begin{bmatrix} \underline{\nu} \end{bmatrix} \tag{4.5}$$

and:

$$\begin{bmatrix} \underline{e}^* \\ \hline -\underline{\lambda}^* \end{bmatrix} = \begin{bmatrix} \underline{e} \end{bmatrix}$$

On the basis of (4.4.) the equilibrium value of $\underline{\lambda}^*$ can be directly determined by substituting $\underline{\lambda}^*$ from the right-hand side into the left-hand side of (4.4.).

Hence it is clear that, once a set of active constraints at the equilibrium point has been identified, a convergent solution is guaranteed, since (4.4.) gives always a unique solution.

In general, however, there is no <u>a priori</u> information concerning the question as to whether at the optimum a certain constraint is active. This implies that during each step of the iteration procedure a set of corner points has to be considered (which is obviously inherent to solution algorithms of L.P. models).

As long as the ultimate set of <u>active</u> constraints has not been identified, a switch among active constraints during each step of the iteration is possible. In the case of linear models and of single conditions for the effluent standards (i.e., no synergetic effects), an equilibrium solution is in general guaranteed, as will be shown now.

Assume, without loss of generality, that there are only inequality conditions in the form of effluent standards. Matrix C is assumed to include also the input-output model from section IV, so that the production variables are always positive; hence there is only a possibility of switches among active constraints for the effluent standards. This would imply the following side-conditions for the primal L.P. program:

$$\left.\begin{array}{l} C\underline{x} = \underline{c} \\ \hat{r}\underline{p} \leq \hat{r}\underline{p}^* = \underline{\bar{p}} \end{array}\right\} \tag{4.7}$$

where $\hat{r}$ is a diagonal scale matrix with positive elements which are set greater than 1. Obviously, the pollutant vector $\underline{p}$ is a sub-vector of $\underline{x}$; it is supposed to form the last elements of $\underline{x}$. In a more concise way, (4.7.) can be written as:

$$\begin{bmatrix} C & \\ \hline 0 & \hat{r} \end{bmatrix} \begin{bmatrix} \underline{x} \end{bmatrix} \begin{array}{c} = \\ \\ \leq \end{array} \begin{bmatrix} \underline{c} \\ \hline \underline{\bar{p}} \end{bmatrix} \tag{4.8}$$

where the lower coefficient matrix is structured according to the rank order of $\underline{p}$ in $\underline{x}$.

It can be directly derived that the general dual constraint set can now be written as:

$$\begin{bmatrix} C' & \begin{array}{c} 0 \\ \hline \hat{r} \end{array} \end{bmatrix} \begin{bmatrix} \underline{\mu} \\ \\ \underline{\lambda}_i \end{bmatrix} \geq \begin{bmatrix} \underline{e}^* \\ \hline -\underline{\lambda}_{i-1} \end{bmatrix} \tag{4.9}$$

where $\underline{\lambda}_i$ represents the shadow price of effluent standards from the ith iteration procedure. In the equilibrium, the difference between $\underline{\lambda}_i$ and $\underline{\lambda}_{i-1}$ should be infinitely small.

As set out above, in the optimum only a square system of linearly independent constraints is relevant (cf. (4.2.) - (4.4.)). Therefore, in a way analogous to (4.4.) a similar system of dual equations can be derived.

Now the recursive procedure proceeds as follows. During the first step of the iteration procedure $\underline{\lambda}_{i-1} = \underline{\lambda}_0 = \underline{0}$. Then $\underline{\lambda}_1$ can be directly calculated in a way similar to (4.4.) by means of a reduced coefficient matrix:

$$\begin{bmatrix} \underset{\cdot}{C}' & \begin{array}{c} 0 \\ \hline \hat{r}_1^* \end{array} \end{bmatrix} \begin{bmatrix} \underline{\mu} \\ \hline \underline{\lambda}_1^* \end{bmatrix} = \begin{bmatrix} \underline{e}^* \\ \hline \underline{0} \end{bmatrix}$$

where $\underline{\lambda}_1^*$ is a subvector of $\underline{\lambda}_1$ the elements of which correspond to active effluent standards, and where r_1^* is reduced accordingly. In order to arrive at a diagonal matrix r_1^* at the right lower corner of the coefficient matrix, the rows of the coefficient matrix C' and of the coefficient vector $\underline{e}^*$ have to be adapted accordingly.

On the basis of (4.10.) $\underline{\lambda}_1^*$ can be calculated

$$\begin{bmatrix} \underline{\mu} \\ \hline \underline{\lambda}_1^* \end{bmatrix} = \left[\begin{array}{c|c} C' & 0 \\ \hline & \hat{r}_1 \end{array}\right]^{-1} \begin{bmatrix} \underline{e}^* \\ \hline \underline{0} \end{bmatrix} \tag{4.11}$$

The inverse coefficient matrix can be calculated by partitioning this matrix as follows:

$$\left[\begin{array}{c|c} C' & 0 \\ \hline & \hat{r}_1^* \end{array}\right] = \left[\begin{array}{c|c} C_1 & 0 \\ \hline C_2 & \hat{r}_1^* \end{array}\right] \tag{4.12}$$

Now it can directly be checked that the inverse matrix of (4.12.) is equal to:

$$\left[\begin{array}{c|c} C_1^{-1} & 0 \\ \hline -(\hat{r}_1^*)^{-1} \; C_2 C_1^{-1} & (\hat{r}_1^*)^{-1} \end{array}\right] \tag{4.13}$$

Therefore, $\underline{\lambda}_1^*$ is equal to:

$$\begin{aligned} \underline{\lambda}_1^* &= -(\hat{r}_1^*)^{-1} \; C_2 C_1^{-1} \; \underline{e}^* , \\ &= (\hat{r}_1^*)^{-1} \; \underline{\pi} \end{aligned} \tag{4.14}$$

where $\underline{\pi}$ is defined as:

$$\underline{\pi} = -C_2 C_1^{-1} \; \underline{e}^* \tag{4.15}$$

Next, the second-phase value of $\underline{\lambda}$, $\underline{\lambda}_2{}^*$, can be calculated in a similar way by means of (4.9.) and (4.10.), so that:

$$\begin{bmatrix} \underline{\mu} \\ \hline \underline{\lambda}_2{}^* \end{bmatrix} = \left[\begin{array}{c|c} C' & 0 \\ \hline & \hat{r}_2{}^* \end{array}\right]^{-1} \begin{bmatrix} \underline{e}^* \\ \hline -\underline{\lambda}_1^* \end{bmatrix} \tag{4.16}$$

Should there be a switch among active constraints, then the rows of coefficients should be interchanged accordingly. It is easily seen that $\underline{\lambda}_2{}^*$ is equal to:

$$\underline{\lambda}_2^* = -(\hat{r}_2^*)^{-1} C_2 C_1^{-1} \underline{e}^* - (\hat{r}_2^*)^{-1} \underline{\lambda}_1^*$$

$$= (\hat{r}_2^*)^{-1} \underline{\pi} - (\hat{r}_2^*)^{-1} (\hat{r}_1^*)^{-1} \underline{\pi} \qquad (4.17)$$

$$= \left\{(\hat{r}_2^*)^{-1} - (\hat{r}_2^*)^{-1} (\hat{r}_1^*)^{-1}\right\} \underline{\pi}$$

where use is made of (4.13.) - (4.15.).

For higher round values of $\underline{\lambda}$ the same procedure can be employed. It is easy to see that the matrix series in this recursive procedure is a convergent series, since all elements of the diagonal matrices $(\hat{r}_1^*)^{-1}$ are (in absolute value) less than 1. The equilibrium value of $\underline{\lambda}$ is, given the L.P. approach, a non-negative vector, which reflects the implicit marginal social costs of imposing effluent standards. This completes the proof that the aforementioned recursive procedure does lead to a convergent equilibrium value of $\underline{\lambda}$.

Finally, attention will be paid to the situation where the product variables $\underline{q}$ may also be equal to zero (for example, in case of absence of an input-output model). An illustration of this case is the following simplified model:

$$\left.\begin{array}{l} \max = \underline{v}'\underline{q} - \underline{\lambda}'\underline{p} \\ \text{s.t.} \\ p = B\underline{q} \\ \hat{r}\underline{p} \leq \underline{p} \\ \hat{t}\underline{q} \geq \underline{0} \quad , \end{array}\right\} \qquad (4.18)$$

where B is a pollution input-output matrix, and $\hat{t}$ a diagonal scale matrix with known elements greater than 1. The primal constraint set can now be written as:

$$\begin{bmatrix} I & -B \\ \hat{r} & 0 \\ 0 & -\hat{t} \end{bmatrix} \begin{bmatrix} \underline{p} \\ \underline{q} \end{bmatrix} \begin{array}{c} = \\ \leq \end{array} \begin{bmatrix} \underline{0} \\ \underline{p} \\ \underline{0} \end{bmatrix} \qquad (4.19)$$

By defining a diagonal matrix $\hat{u}$ as:

$$\hat{u} = \begin{bmatrix} \hat{r} & 0 \\ 0 & -\hat{t} \end{bmatrix} \qquad (4.20)$$

the primal constraint set can be transformed into a dual constraint set in a way analogous to (4.9.):

$$\begin{bmatrix} I & \\ \hline -B' & \hat{u} \end{bmatrix} \begin{bmatrix} \underline{\mu} \\ \\ \underline{\lambda}_i \end{bmatrix} = \begin{bmatrix} \underline{v} \\ \\ -\underline{\lambda}_{i-1} \end{bmatrix} \qquad (4.21)$$

where $\underline{\mu}$ represents the shadow prices associated with the equality constraints and $\underline{\lambda}$ the shadow prices associated with the inequality constraints. It should be noted that the eigenvalues of the matrix $\hat{u}$ exceed 1.

The dual constraint set related to active constraints can be written as:

$$\begin{bmatrix} I & 0 \\ \hline -B' & \hat{u}^* \end{bmatrix} \begin{bmatrix} \underline{\mu} \\ \hline \underline{\lambda}_i^* \end{bmatrix} = \begin{bmatrix} \underline{v} \\ \hline -\underline{\lambda}_{i-1}^* \end{bmatrix} \qquad (4.22)$$

where $\hat{u}^*$ and $\underline{\lambda}^*$ are again the remaining parts of $\hat{u}$ and $\underline{\lambda}_i$ associated with active constraints (after a permutation of the rows of the coefficient matrix).

The structure of (4.22.) is completely identical to that of (4.12.), so that in this case also a convergent solution is guaranteed.

The previous method will now be illustrated by means of a numerical example (section V), while next an empirical application will be presented (section VI).

V A NUMERICAL ILLUSTRATION

In this section an illustration will be presented of the method described above by employing shadow prices as a means for determining the social costs caused by imposing effluent standards. Obviously, a rise in the permissible level of air pollution will lead to an increase of the corrected regional income surplus. Given this permissible level of income, a numerical calculation of the implicit evaluation of the environmental quality can be derived as follows.

Suppose the permissible level of discharges of pollutant 1 is p_1^*. If this effluent constraint is active, a marginal decrease of p_1^* the size of dp_1 would lead to a decline in income. Given the fact, that the permissible level of air pollution is set equal to p_1^* and not equal to $p_1^* - dp_1$, one may draw the conclusion that, apparently, from p_1^* downwards one additional unit of increased air quality is lower evaluated than a marginal income decline. Hence, the value of environmental commodities, which would have been obtained in consequence of an additional unit decline of pollutant 1, is implicitly equal to (or lower than) the resulting income effect.

Obviously, such an implicit evaluation of environmental commodities can only be applied to a marginal shift near the optimum. In order to assess the total damage, one may assume that the foregoing marginal evaluation procedure can be applied to all levels of pollutant 1, so that an aggregation of marginal costs will lead to an environmental cost curve. This idea will be illustrated now by means of a numerical example.

Suppose the following model:

$$\left\{ \begin{array}{l} \max \omega = [2\ 3] \begin{bmatrix} x_1 \\ x_2 \end{bmatrix} - [\lambda_1\ \lambda_2] \begin{bmatrix} p_1 \\ p_2 \end{bmatrix} \\ \text{s.t.} \\ \text{(i)} \quad p_1 \leq p_1^* \qquad \text{(iii)} \quad p_1^* = 5 \qquad \text{(v)} \quad \begin{bmatrix} p_1 \\ p_2 \end{bmatrix} = \begin{bmatrix} 1 & 2 \\ 3 & 2 \end{bmatrix} \begin{bmatrix} x_1 \\ x_2 \end{bmatrix} \\ \text{(ii)} \quad p_2 \leq p_2^* \qquad \text{(iv)} \quad p_2^* = 30 \end{array} \right. \tag{5.1}$$

where variables x_1 and x_2 are the production levels of good 1 and 2, parameters 2 and 3 (out of the objective function) the corresponding value added coefficients, p_1 and p_2 the effluent discharges of pollutant 1 and 2, p_1^* and p_2^* the effluent standards for pollutant 1 and 2, λ_1 and λ_2 the implicit shadow costs per unit of pollutant 1 and 2, and where the elements of the matrix in (v) represent the pollution-output coefficients, respectively.

By substituting (v) into ω, and (i) and (ii), (5.1.) can be written as:

$$\left. \begin{array}{l} \max \omega = [2\ \ 3] \begin{bmatrix} x_1 \\ x_2 \end{bmatrix} - [\lambda_1\ \lambda_2] \begin{bmatrix} 1 & 2 \\ 3 & 2 \end{bmatrix} \begin{bmatrix} x_1 \\ x_2 \end{bmatrix} \\ \text{s.t.} \\ \begin{bmatrix} 1 & 2 \\ 3 & 2 \end{bmatrix} \begin{bmatrix} x_1 \\ x_2 \end{bmatrix} \leq \begin{bmatrix} 5 \\ 30 \end{bmatrix} \end{array} \right\} \tag{5.2}$$

The crucial element of the solution procedure is to determine for the primal program those values for λ_1 and λ_2 that, at the optimum, they correspond to the shadow prices of the dual L.P. program, so that the shadow prices of the dual are again λ_1 and λ_2. The essential assumption on determining λ_1 and λ_2 is that, at the optimum, the total damage costs are calculated on the basis of aggregated marginal shadow costs. In our example, the values of λ_1 and λ_2 (calculated by means of the recursive technique described above) are respectively equal to 1 and 0, whereas the optimal values of x_1 and x_2 are equal to 5 and 0, respectively.

The dual model associated with (4.2.) can be constructed as:

$$\left.\begin{array}{l} \min \phi = [5 \quad 30] \cdot \begin{bmatrix} y_1 \\ y_2 \end{bmatrix} \\ \text{s.t.} \\ \begin{bmatrix} 1 & 3 \\ 2 & 2 \end{bmatrix} \begin{bmatrix} y_1 \\ y_2 \end{bmatrix} \geq \begin{bmatrix} 1 \\ 1 \end{bmatrix} \end{array}\right\} \quad (5.3)$$

Obviously the first constraint of the dual model is active due to the fact that $x_1 = 5(\neq 0)$. Furthermore, only the first constraint of the primal model is active, so that one may conclude that $y_1 \neq 0$, $y_2 = 0$. Therefore, the dual constraint set can be written as:

$$\left.\begin{array}{r} y_1 + 3y_2 = 1 \\ y_2 = 0 \end{array}\right\} \quad (5.4)$$

from which the equilibrium values of the dual variables can be calculated: $y_1 = 1$ and $y_2 = 0$; y_1 and y_2 are the shadow prices, which are obviously equal to λ_1 and λ_2. Hence, $\lambda_1 = 1$ and $\lambda_2 = 0$. which was the result described above.

The next step to be undertaken is to detect the limits between which p_1^* can vary without changing the basic variables. For that purpose, a sensitivity analysis is carried out (see Zionts [1974]). If the value of p_1^* falls between 0 and 10, the same variables appear to remain basic (suppose all other things remain constant). Therefore, over the whole range: $0 \leq p_1 \leq 10$, the shadow prices λ_1 and λ_2 are constant and equal to 1 and 0, respectively.

The next step is to analyse the situation where $p_1^* \geq 10$. Obviously, there will be a change in the basic variables, so that probably some other constraints will become active. This new problem can be formulated as:

$$\left\{\begin{array}{l} \max \omega = [2 \quad 3] \begin{bmatrix} x_1 \\ x_2 \end{bmatrix} - [\lambda_1^* \quad \lambda_2^*] \begin{bmatrix} p_1 - 10 \\ p_2 \end{bmatrix} - 10.1 \\ \text{s.t.} \\ \text{(i)} \ p_1 \leq p_1^* \quad \text{(iii)} \ p_1^* = 12 \quad \text{(v)} \begin{bmatrix} p_1 \\ p_2 \end{bmatrix} = \begin{bmatrix} 1 & 2 \\ 3 & 2 \end{bmatrix} \begin{bmatrix} x_1 \\ x_2 \end{bmatrix} \text{(vi)} \ p_1 \geq 10 \\ \text{(ii)} \ p_2 \leq p_2^* \quad \text{(iv)} \ p_2^* = 30 \end{array}\right. \quad (5.5)$$

The objective function has been changed in order to take account of the fact that the quantity of pollutant 1 until 10 units causes a damage of $10\lambda_1 = 10 \times 1$ monetary units <u>at most</u>. For all quantitites of pollutant 1 above 10 units a new shadow price has to be calculated.

The same procedure as described above can be carried out to find the new corresponding shadow prices λ_1^* and λ_2^*. In this situation their values appear to be equal to 0.625 and 0.125, respectively.

On the basis of a new sensitivity analysis, it turned out that at the domain: $10 \leq p_1^* \leq 30$ the same variables remained in the basis. Next, for $p_1 \geq 30$, the problem can be reformulated as:

$$\left\{\begin{array}{l} \max \omega = [2 \quad 3] \begin{bmatrix} x_1 \\ \\ x_2 \end{bmatrix} - [\lambda_1^{**} \ \lambda_2^{**}] \begin{bmatrix} p_1 - 30 \\ \\ p_2 \end{bmatrix} - 10\lambda_1 - 20\lambda_1^* \\ \text{s.t.} \\ \text{(i)} \ p_1 \leq p_1^* \quad \text{(iii)} \ p_1^* = 32 \quad \text{(v)} \begin{bmatrix} p_1 \\ p_2 \end{bmatrix} = \begin{bmatrix} 1 & 2 \\ 3 & 2 \end{bmatrix} \begin{bmatrix} x_1 \\ x_2 \end{bmatrix} \quad \text{(vi)} \ p_1 \geq 30 \\ \text{(ii)} \ p_2 \leq p_2^* \quad \text{(iv)} \ p_2^* = 30 \end{array}\right. \qquad (5.6)$$

The solution for λ_1^{**} and λ_2^{**} appears to become: $\lambda_1^{**} = 0$ and $\lambda_2^{**} = 0.75$. The corresponding sensitivity analysis showed that for $p_1^* \geq 30$ the same variables are always basic.

Summarizing, the following scheme can be constructed:

Table 1.

Summary of computational results

$p_2^* = 30$	$0 \leq p_1 \leq 10$	$\lambda_1 = 1$	$\lambda_2 = 0$	$x_1 \neq 0$	$x_2 = 0$
$p_2^* = 30$	$10 \leq p_1 \leq 30$	$\lambda_1^* = 0.625$	$\lambda_2^* = 0.125$	$x_1 \neq 0$	$x_2 \neq 0$
$p_2^* = 30$	$p_1 \geq 30$	$\lambda_1^{**} = 0$	$\lambda_2^{**} = 0.75$	$x_1 = 0$	$x_2 \neq 0$

This scheme can be transformed into the following figure where a relation between the total loss of income (and, implicitly, the evaluation of environmental damage) and the level of p_1 is given (with a constant level of p_2 equal to 30).

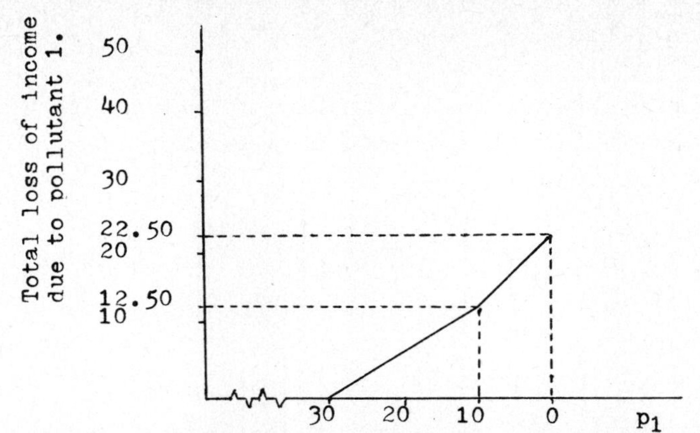

Figure 2. Figurative representation of computational results.

VI APPLICATION

The method of determining the implicit social costs of effluent standards has been applied to a newly created industrial area in the Netherlands, viz. the Meuse-flat (Massvlakte). This area has been recaptured from the North Sea, and it serves to extend the industrial basis of the Rhine Delta area around Rotterdam.

The major problem was formed by the question: which industrial activities should be located at the Meuse-flat? Various meaningful combinations of candidate-activities have to be inspected in order to arrive at a definite conclusion as to the optimal mix of activities. This zero-one selection problem is the subject of another study carried out by one of the authors.

Here, only one of the meaningful combinations of candidate-activities for locating at the Meuse-flat will be considered. Given such a combination of activities, a L.P. model will be used to calculate the optimal land use requirements of these activities, the optimal level of pollution permitted, as well as the regional shadow costs of imposing effluent standards.

The combination of activities considered here includes: iron and steel works, petro-chemical activities, tanker-cleaning and reparation, container terminal, and transhipment activities. The optimal level of these activities is determined by means of a regional efficiency criterion which aims at maximizing net regional benefits. The constraints are among others effluent standards for 2 types of pollutants (viz. sulfur dioxide and particulates), and land use constraints for the activities. Therefore, in a manner analogous to sections III and IV the following program can be formulated:

$$\left.\begin{aligned} &\max\ \omega = \underline{u}'\underline{x} - \underline{\lambda}'p \\ &\text{s.t.} \\ &\underline{p} = B\ \underline{x} \\ &\underline{p} \leq \underline{p}^* \\ &\underline{i}'\underline{x} \leq x \\ &\underline{x} \leq \underline{x}^o \end{aligned}\right\} \qquad (6.1)$$

where $\underline{x}$ is an (unknown) vector of land use requirements per activity, $\underline{u}$ a (known) vector of value added coefficients per unit of land use for each activity, B a matrix with emission coefficients of pollution from the various activities, $\underline{p}$ the unknown vector of pollutants with effluent standard $\underline{p}^*$, x the total area available for industrial activities, and $\underline{x}^o$ the maximum quantity of land available for each individual activity.

The data of this model are contained in table 2.[1] They are based on the most recent information available.

Table 2

Data of the pollution model

	$\underline{u}$	B'		$\underline{x}^o$
Iron and Steel Works	183	48	34	500
Petro Chemical Activities	62	48	0	40
Tanker Cleaning & Reparation	51	0	0	100
Container Terminal	79	42	1	150
Trans-shipment Activities	112	0	0	15

	$\underline{p}^*$
Particulates	1161
SO_2	1948

$x^* = 780$

where $\underline{u}$ is measured in 1000 Dfl.; b_{ij} in tons per year per ha; $\underline{x}^o$ and x^* in ha, and $\underline{p}^*$ in tons per year.

By employing the method described in sections III and V, the optimal results can be calculated. These results are presented in table 3.

Table 3

Results of the pollution model

$p_2^* = 1948$	$0 \leq p_1 \leq 2750$	$\lambda_1 = 1.91$	$\lambda_2 = 0$
$p_2^* = 1948$	$2750 \leq p_1 \leq 8838$	$\lambda_1 = 0.91$	$\lambda_2 = 1.41$

The curve representing the cost function associated with environmental constraints is represented in Figure 3., where a range is represented for the effluent standards of particulates.

---

[1]The authors wish to thank Ad van Delft for his assistance in collecting the data.

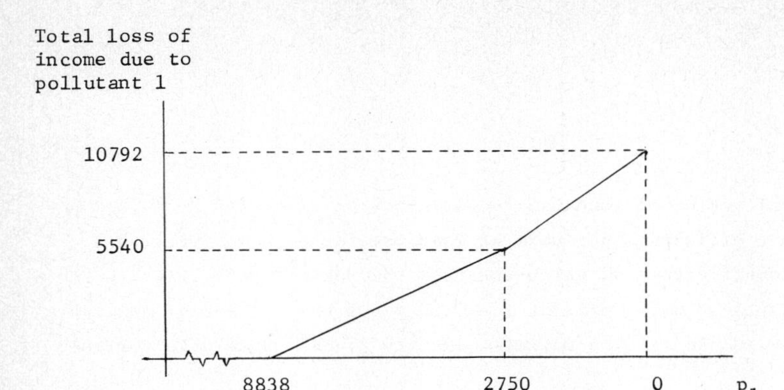

Figure 3. Diagrammatic representation of computation results

What conclusion can be drawn from Table 3? Given the L.P. approach, given the assumption that the (maximal) damage can be measured by means of the shadow price via a linear function, and given the selected activities and the levels of effluent standards, one may conclude that the loss of income per unit of a downward shift in $p^*_{.}$ (given a constant $p_2^* = 1948$), is equal to 0.91 (at the range from 8838 to 2750) and equal to 1.91 (at the range from 2750 downwards). These figures will obviously change when the given value of p_2^* is changed, or other activities are selected as possibilities for locating at the Meuse-flat.

Therefore, it is necessary to be aware of the restricted assumptions underlying these figures. The above mentioned procedure is just a first attempt to find a possibility that enables us to quantify the damage caused by air pollution and other environmental damages.

VII CONCLUSION

In the previous paragraphs a method has been developed to gauge the social costs of preventing environmental damage by imposing effluent standards. The crucial element in this approach was the fact that these costs played an essential role in determining an optimal pattern of production and of pollution.

The fundamental reason for the necessity of using these rather complicated methods stems from the multiplicity of phenomena to be considered in public decision-making. The use of a single income criterion as a measure for optimal public decisions would neglect other relevant policy variables like (un)employment and pollution. The basic question is thus: how to integrate different variables (with different dimensions) into one decision framework?

In this respect, it may be worthwhile to refer to several methods recently developed to deal with multiple criteria in regional planning (for a survey, see Nijkamp [1974a]). These multi-criteria decision problems may be illustrated by transforming the pollution model from section VI into the following vector maximization problem (cf. Kornbluth [1974], Philip [197]] and Zeleny [1974]):

$$\left.\begin{array}{l} \left\{\max \quad \omega_0 = \underline{u}'\underline{x}\right. \\ \left\{\begin{array}{lll} \min & \omega_1 & = p_1 \\ - & - & - \\ - & - & - \\ \min & \omega_K & = p_K \end{array}\right. \\ \text{s.t.} \\ \underline{p} = B\ \underline{x} \\ \underline{i}'\underline{x} = x^o \\ \underline{x} \leq \underline{x}^o \end{array}\right\} \qquad (7.1)$$

The previous model is essentially based on several performance measures. It is clear, that a simultaneous optimization of all objective functions will, in general, not lead to a single optimal solution to the problem. However, it is possible to define an efficient (or Pareto-optimal) solution to the foregoing multi-objective L.P. models. A solution is efficient or non-dominated, if it is not possible to increase one of the objective functions without decreasing one of the others.

Such a multi-objective approach leads, in general, to a set of efficient solutions, which can be considered as 'reasonable' or 'satisficing' solutions. Next, it is the decision-maker's responsibility to select an ultimate solution out of the set of efficient solutions.

The foregoing multi-objective approach has not yet been applied to the decision problem specified in section VI, since so far only a few steps have been undertaken to develop adequate solution algorithms (although progress is being made at the moment). It is the opinion of the authors, that,in the future, multi-objective methods will be promising tools for regional planning and decision-making.

REFERENCES

Bain, J.S., 1973. Environmental Decay, Little, Brown and Company, Boston.

Barrett, L.B. and Waddell, T.E., 1970. "The Cost of Air Pollution Damages: A Status Report", U.S. Department of Health, Education and Welfare; Environmental Protection Agency.

Burton, E.S. and W. Sanjour, "A Simulation Approach to Air Pollution Abatement Program Planning", Socio-Economic Planning Sciences, vol. 4, March. 1970, pp. 147-159.

Cantilli, E.J. Programming Environmental Improvements in Public Transportation, Heath, Lexington (Massachusetts), 1974.

Coase, R.H., "The Problem of Social Cost", The Journal of Law and Economics, vol. 3, 1960, pp. 1-44.

Crocker, T.D., Urban Air Pollution Damage Functions, University of California, Riverside, 1972.

Dales, J.H., Pollution, Property and Prices, University of Toronto Press, Toronto, 1968.

ENVIRONNEMENT, Evaluation du Coût de la Prévention de la Pollution Atmosphérique dans l'Industrie en France, La Documentation Francaise, Paris, 1974.

Freeman III, A.M., R.H. Haveman and A.V. Kneese, The Economics of Environmental Policy, Wiley, New York, 1973.

Goudzwaard, B., Ongeprijsde Schaarste, Van Stockum, The Hague, 1970.

Hueting, R., Nieuwe Schaarste en Economische Groei, Agon-Elsevier, Amsterdam, 1974.

Isard, W., Ecologic-Economic Analysis for Regional Development, Free Press, New York, 1972.

Jaksch, J.A. and H.H. Stoevener, "Effects of Air Pollution on Residential Property Values in Toledo, Oregon", Report 304, Agricultural Experiment Station, Oregon State University, 1970.

Jansen, H.M.A., et al., Een Raming van Schade door Luchtverontreiniging in Nederland in 1970, Instituut voor Milieuvraagstukken, Vrije Universiteit, Amsterdam, 1974.

Klaassen, L.H., and A.C.P. Verster, Kosten-baten analyse in Regionaal Perspectief, Tjeenk Willink, Groningen, 1974.

Klarmen, H.E., The Economics of Health, Columbia University Press, New York, 1965.

Kneese, A.V., "Rationalizing Decisions in the Quality Management of Water Supply in Urban-Industrial Areas", The Public Economy of Urban Communities (J. Margolis, ed.), Johns Hopkins Press, Baltimore, 1965, pp. 170-191.

Kohn, R.E., "Linear Programming Model for Air Pollution Control", Journal of the Air Pollution Control Association, vol. 20, no. 2, 1970, pp. 78-82.

Kornbluth, J.S.H., "Duality, Indifference and Sensitivity Analysis in Multiple Objective Linear Programming", Operational Research Quarterly, vol. 25, no. 4, 1974, pp. 599-614.

Lave, L.B. and E.P. Seskin, "Health and Air Pollution", Swedish Journal of Economics, vol. 109, no. 1-2, 1971, pp. 76-95.

Leontief, W., "Environmental Repercussions and the Economic Strucutre: An Input-Output Approach", Review of Economics and Statistics, vol. 52,1970, pp. 262-271.

Little, I.M.D., A Critique of Welfare Economics, Clarendon Press, Oxford, 1950.

Mastenbroek, A.P. and P. Nijkamp, "Een Multi-regional Milieu-model voor een Ruimtelijke Allocatie van Investeringen", Milieu en Economie (P. Nijkamp, ed.), Rotterdam University Press, Rotterdam, 1974.

Mishan, E.J., The Costs of Economic Growth, Allen and Unwin, London, 1967.

Mishan, E.J., Cost Benefit Analysis, Allen & Unwin, London, 1971.

Musgrave, R.A., Fiscal Systems, Yale University Press, New Haven, 1969.

Nijkamp, P. and W.H. Somermeijer, "Explicating Implicit Social Preference Functions", The Economics of Planning, vol. 11, no. 3, 1971, pp. 101-119.

Nijkamp, P. and J.H.P. Paelinck, "Some Models for the Economic Evaluation of the Environment", Regional and Urban Economics, vol. 3, no. 1, 1973, pp. 33-62.

Nijkamp, P. and J.H.P. Paelinck, "An Interregional Model of Environmental Choice", Papers of the Regional Science Association, vol. 31, 1973, pp. 51-71.

Nijkamp, P., "Spatial Interdependencies and Environmental Effects", Dynamic Allocation of Urban Space, (A. Karlqvist, L. Lundqvist and F. Snickars, eds.), Saxon House, London, 1974, pp. 175-209.

Nijkamp, P., "A Multi-criteria Analysis for Project Evaluation", Paper presented at the North-American Meeting of the Regional Science Association, Chicago, 1974 (to be published in the Papers of the Regional Science Association).

Nijkamp, P. and J.H.P. Paelinck, Operational Theories and Methods in Regional Economics, Saxon House, London, 1975 (forthcoming).

Opschoor, J.B., Economische Waardering van Milieuverontreiniging, Van Gorcum, Assen, 1974.

Pasquill, F., Atmospheric Diffusion, D. van Nostrand Co., London, 1962.

Philip, J., "Algorithms for the Vector Maximization Problem", Mathematical Programming, vol. 2, no. 3, 1972, pp. 207-224.

Pigou, A.C., The Economics of Welfare, MacMillan, London, 1920.

Plowden, S.P.C., The Cost of Noise, Metra, London, 1970.

Ridker, R.R., "Strategies for Measuring the Cost of Air Pollution", The Economics of Air Pollution, A Symposium, (H. Wolozin, ed.), Norton, New York, 1966, pp. 87-101.

Ridker, R.R., Economic Costs of Air Pollution, Praeger, New York, 1967.

Ridker, R.G. and J.A. Henning, "The Determinants of Residential Property Values with Special Reference to Air Pollution", Review of Economics and Statistics, vol. XLIX, no. 2, 1967, pp. 246-257.

Ridker, R.G., "Standard-setting as a Frame of Reference", Economic Thinking and Pollution Problems (D.A.L. Auld, ed.), University of Toronto Press, Toronto, 1972, pp. 22-29.

Russell, C.S. and W.O. Spofford, "A Quantitative Framework for Residuals Management Decisions", Environmental Quality Analysis (A.V. Kneese and B.T. Bower, eds.), Johns Hopkins Press, Baltimore, 1972, pp. 115-179.

Stratford, A.H., Airports and the Environment, MacMillan, London, 1974.

Victor, P.A., Economics of Pollution, MacMillan, London, 1972.

Wolozin, H. (ed.), The Economics of Air Pollution: A Symposium, Norton, New York, 1966.

Zeleny, M., Linear Multiobjective Programming, Springer, Berlin, 1974.

Zionts, S., Linear and Integer Programming, Prentice Hall, Englewood Cliffs, New Jersey, 1974.

THE RISKS AND BENEFITS FROM LEAD IN GASOLINE: EFFECTS ON ENERGY USE AND ENVIRONMENT

Robert W. Resek and George Provenzano

I INTRODUCTION

The University of Illinois Institute for Environmental Studies has been examining environmental pollution by lead and other metals in a coordinated interdisciplinary study since 1970. The major efforts in the study have been by biologists, agronomists, chemists, ecologists, etc., who have reached many specific, useful conclusions concerning the distribution, movement and effects of lead in the environment. (Metals Task Force, 1974a and 1974b; Miller, et al., 1974; and references cited therein.)

In this paper, we contrast, in a preliminary way, the environmental risks and economic benefits of utilizing lead as an additive in automotive fuel. In this analysis some of the environmental and economic effects are found to be large, and some are small, but in the analytical process we depend on the prior knowledge gained from interdisciplinary research, and assume that these results are employed in order to minimize the potential damages from lead pollution. Thus, the research has already had a substantial payoff in enabling farmers, etc., to minimize any lead effects.

In this work, three major aspects are considered. The first two are the potential negative environmental impacts of automotive lead emissions. These include (1) the risk of damage to agriculture and (2) the risk of damage to public health. Finally, we consider (3) the benefits of having lead in gasoline.

II AGRICULTURAL TASK

Agricultural risk from lead pollution may initially be divided into two portions. First is the loss from either reduced yield (output) or increased agricultural expense to avoid reduced yield. Second is the risk to human health from the presence of lead in or on the food source.

II.1 Reduced Food Output

This analysis requires consideration of several component parts. First, we discuss the amount of lead that is transmitted to plants from automobile emissions. Next, we consider the effect on plants of being grown where there is lead in the soil or in the air. Finally, we assess the dollar value of the risk of crop exposure to lead.

It should be apparent that each element of this analysis rests on a very substantial research base developed by the appropriate physical, chemical, and/or biological scientists. Research results from these studies provide essential inputs into the economist's analysis of crop losses due to lead pollution. Because the studies are still underway, the statements here (including economic benefits) must be preliminary. The explanations are very brief and numerically some approximations are made. Nevertheless, the order of magnitude is correct.

Most of the lead emitted from automobiles settles out along the roadside in a band that is only a few meters wide. Furthermore, lead accumulates within this band because of the chemical binding capacity of Illinois soils; the movement of lead in soils occurs primarily through physical means, e.g., water and wind erosion.

The concentration of lead in soil as a function of traffic density and distance from the road was determined by analyzing soil samples collected along transects perpendicular to the highways. This analysis indicated a background level of 15 ppm (parts-per-million) lead in the soil throughout central Illinois; this is lead which cannot be attributed to automotive lead emissions.

For this discussion we classify traffic as "heavy" if there are 10,000 vehicles per day, as "moderate" for 2,500, and as "light" for 200 or less vehicles per day. These measures of traffic density roughly correspond to a 4-lane expressway, a two-lane through road, and a local road serving farm residences respectively. Throughout Illinois there is a grid of township roads 1 mile apart (1.6 km.) with "light traffic." In addition, there is a "moderate traffic" road about every 15 miles (25 km.) and a "heavy traffic" road about every 100 miles (160 km.).

Along the heavy traffic roads soil samples show a very high lead level adjacent to the road surfaces (.3 meter). Lead concentrations decrease geometrically until the background level is reached at around 50 meters from the road. For a heavy traffic road, the lead level at 10 meters is 70 ppm while for a light traffic road it is 20 ppm at this distance. The mean level in the first 50 meters may be estimated as 45 ppm and 18ppm for these two road types. The use of these figures will likely overstate the danger of damages from lead because the first 5 to 20 meters along the roadside are not farmed in many cases.

The effect of continued lead emissions is cumulative; lead concentrations in soil will increase as more is added to present levels. Traffic growth in the area studied has averaged 3% a year. To provide a rough estimate of the increase in lead, we assume traffic has grown geometrically in the past and will continue to do so in the future. We then solve for the number of years required to double lead levels. The reader may do this for himself. If the annual growth rate is p, the concentration of lead will double (over background) in n = ln 2/p = .7/p years. If we assume p is 5% (to shorten the time and increase the effect of lead) we find n = 14. Thus, at 10 meters from a heavy traffic road, lead level in 14 years will be 15 + 2(45) = 105 ppm, in 28 years will be 195 ppm, and in 42 years will be 375 ppm. These values are approximate but show the potentially obtainable levels.

Within the 50 meter band adjacent to the roads, lead levels vary significantly. However, we measure this area as that primarily susceptible to automotive lead emissions. If we consider the 50 meter band and the road grid discussed above, then 12% of all land in Illinois is within this distance along light traffic roads; .8% near moderate traffic roads; and .01% along heavy traffic roads. These factors are necessary in assessing the total impact of lead emissions.

The effect of soil lead on plant growth is now considered. For this, plants were grown in pots with varying lead content. There was a major difference in the results depending on the type of soil employed in the experiment. For nearly any type of soil, if lead approaches 10,000 ppm, there is a major effect on the growth of the plant. However, for lead levels comparable to those found along highways, a significant effect on plant growth occurred in only two soil types--sandy soils and acidic soils. Sandy soils make up 3.4% of Illinois land while potentially acidic soils are found in the southern 1/3 of the state (Agricultural Experiment Station, 1967).

Consider first the acidic soils. Good farming practice (without regard to lead) causes the farmer to regularly spread lime on the fields to reduce the acidity (raise the pH to 7.0). This has been done for many years. At a pH of 7.0 there is essentially no effect from lead at levels considered here.

Even if the lead level should go higher (say to 500) in acidic soils, it is still feasible to eliminate the effects by adding extra lime. This is an example of a case where the risk is not the loss in yield but the extra cost of a newly required treatment. The total needed treatment would be a one-time application of lime (which costs about $20 per acre or $8 per hectare) followed by annual application of micronutrients ($3 to $4 per acre per year or $1.50 per hectare per year).

Illinois has 36,096,000 acres of which 1/3 of .01% or 1,203 acres (2,964 hectares) are acidic and within 50 meters of a heavy traffic road. The cost of treating these soils would be a one-time cost of $24,000 plus $4,500 per year. At a 10% discount rate the one-time cost is $2,400 per year so the total is $6,700 per year for acidic soils after considerable additional lead has accumulated (say year

2015).

In the sandy soils it was found that lead present at 250 ppm causes a growth reduction of 30% in young plants. Illinois has 1,238,000 acres of sandy soils of which 9,904 are by moderate traffic roads and 124 are by heavy traffic roads. We assume, as an initial factor for estimation, that yield loss in the year 2015 will be 30% by moderate traffic and 100% by heavy traffic roads assuming no reduction in lead emissions. Additionally, we assume the full yield is valued at $200/acre or $80/hectare in 1975 prices. The total loss on this basis is $619,000 per year commencing in 2015.

We should reemphasize that these results depend on good agricultural practice which employs in practice the knowledge gained from agronomists' study of lead. In addition, we have tended to overemphasize potential loss and overstate dollar loss. The combined loss for all soils is seen to be around $625,000 per year after 2015. If this analysis is applied to other regions or states, the results could be somewhat different since Illinois has very little sandy soil. The potential problem areas in the United States occur in Nebraska, Wisconsin, and Florida. The latter is the greatest concern because of the juxtaposition of heavy agriculture on sandy soil with heavy vehicle traffic.

II.2 The Risk of Lead in Food

The biologic studies show that to a large degree plants do not take up much lead at concentrations currently found in Illinois soils. In addition, of the amount that is taken up, only a small percentage of the lead is transmitted to the grain or to the edible portions of the plant. Based on the field studies there is less than 1 ppm in the grain.

Another source of food contamination is lead dust that settles on the plant and not in the plant itself. Examples of this type of contamination include lead that settles on leafy vegetables such as lettuce and spinach. Most of this surface lead can be removed by washing. Under usual circumstances, agricultural animals similarly do not seem to take lead into their systems to any great degree according to present research results. We conclude that there appears to be no significant monetary risk from lead that is solely from the food source.

III THE RISK OF LEAD IN THE AIR AND DUST IN URBAN AREAS

Another source of lead that is of potential importance to human health is lead in the urban atmosphere and in urban dusts both in the interior and exterior of urban dwellings. Although it is generally agreed that the main source of airborne lead is the automobile (exceptions occur in the immediate vicinity of primary and secondary lead smelters), there is considerable disagreement over the source of lead in dust. Because lead was a primary constituent of paints which were manufactured before World War II, paint chips and flakes (along with gasoline) may be a major source of lead in dust. This dust is both in the air and on the ground in urban areas and is potentially a danger to humans from being breathed or eaten (as by children at play or by persons who do not adequately wash their hands before eating).

At present chemists are determining the distribution of lead in urban air and dust as well as its original source, e.g. cars, paint. Subsequently, a connection will be made to the dangers of poor health or mortality from the presence of lead from automobiles or other sources.

No specific results are available at this time in a form suitable for risk-benefit analysis. Thus, we will proceed by considering how the results could be stated and the implications of those assumed results.

It is possible that urban lead affects 10 percent of the total population and that the life expectancy of each person is reduced one week for each year of exposure to lead. To evaluate in dollars this loss to society, we need to determine the present value to a person of shortened life span. No solid clear answers are available to this. However, we speculate that a person may be willing to pay $1,000 at age 30 for a one year increase in life expectancy. While one can argue over that value it is difficult to believe it is wrong by a factor of 10.

The total population of the United States was about 203 million and of Illinois was 11 million in 1970. Thus, the loss under these assumptions from lead in urban dust is 203 x 10% x 1/50 x $1000 = $406 million annually for the United States and $22 million annually for Illinois.

The losses determined in this section and above will be contrasted with gains from use of lead below.

IV ENERGY SAVINGS BENEFITS OF LEAD

To fully analyze the potential energy savings, supply and demand relations for gasoline must be precisely known. The supply schedule is subject to great question for the United States and other countries of the world. Additionally, it seems to be predicted on political rather than economic considerations and is extremely difficult to forecast. We consider two major alternatives. First, unlimited quantities of crude oil are available at a fixed price and sufficient refining capacity is also available. Second, a fixed quantity of crude oil comparable to current use is available at a fixed price. Beyond that no additional crude oil may be employed. In principle this could arise because of an export restriction by producing countries, by limited refinery capacity or by an import quota imposed by the U.S. government (which has been suggested).

For any of these possibilities the demand function for gasoline is required. Extensive analysis of demand has been made as part of this project. We feel current estimates by others as well as by ourselves are deficient in that they do not employ data on automobile purchase or stocks. The theory of gasoline demand indicates that a major effect of gasoline prices is on automobile purchases and shows up in gasoline consumption only with a very long lag. Future work in the project will include the estimation of simultaneous demand equations for automobiles and gasoline.

We obtained gasoline demand elasticity from: $\log Q_t = -.14 \log P_t + .47 \log Y_t + .34 \log Q_{t-1} +$ constant, where Q is quantity, P is price, and Y is disposable personal income. This indicates a short run price elasticity of .14 and long run elasticity of .21.

V PHYSICAL EFFECTS OF USING LEAD ADDITIVES IN GASOLINE

Adding tetraethyl lead and other lead antiknock agents to gasoline has provided a dual means of holding down the total energy requirements for automobile transportation. On the one hand, the use of lead additives has enabled petroleum refiners to conserve on energy resources during the production of high-octane gasolines, while on the other hand, high-octane gasolines have permitted improvements in automotive engine fuel economy.

V.1 Petroluem Refining

Gasoline is potentially by far the most valuable product obtained from crude oil, and U.S. refiners have for decades developed processes which have increased the octane rating and output of gasoline per barrel of crude oil. Cracking processes such as hydro, catalytic, and thermal, for example, break heavier petroleum molecules into smaller, more volatile components in the gasoline range, and synthesizing processes such as catalytic reforming and alkylation produce higher octane hydrocarbons for blending.

Octane ratings of refined gasolines are increased further by adding tetraethyl lead. Regular and premium grades of gasoline typically have clear octane ratings of 86-88 and 92-94 RON, respectively. Octane ratings can be increased to 94 and to 100 RON by adding 2.1 - 2.4 grams of lead per gallon to regular and 2.7 - 2.8 grams per gallon to premium. These figures represent a range of average annual concentrations of lead added to gasoline since 1965. The actual amount of lead that is added to gasoline at any one time varies considerably depending on the season of the year and on the region of the country (Bureau of Mines, 1965-72).

The production of high-octane gasoline without lead is more expensive in terms of crude oil and refinery energy requirements. Without lead, refiners must increase clear octane ratings by blending in larger quantities of the more expensive, higher octane hydrocarbons such as branched-chain paraffins and aromatics. In very general terms this means that the amount of alkylation and reforming processing must be increased for a given crude oil throughout. An increase in the intensiveness of these operations results in a smaller output of gasoline per barrel of crude and, in addition, requires larger amount of process heat, steam and electricity.

V.2 Automotive Engine Efficiency

Although several engine specifications--for example, compression ratio, displacement, and air-fuel ratio--influence fuel economy, it is well established that an increase in compression ratio (holding all other specifications constant) will produce better fuel economy. Increases in compression ratio of up to 14:1 will produce increases in an engine's ability to do work with greater efficiency or with greater power per gallon of gasoline consumed.

As compression ratio increases, the corresponding motor fuel octane number must also increase. Octane number is a measure of the ability of a fuel to resist engine knock. Knocking occurs when combustion temperatures cause the air-fuel mixture to ignite prematurely and the presence of knocking indicates inefficient engine operation in terms of fuel consumption. Because combustion temperature and engine efficiency both increase with increases in compression ratio, engine knock also places an upper limit on efficient engine design.

In the 15 years preceding 1971, increases in compression ratio have probably offset losses in fuel economy due to increases in vehicle weight and to the number of installed power accessories. During this period, compression ratios steadily increased on 11 production model cars from a sales-weighted average of 8.13:1 in 1956 to 9.27:1 in 1970 (Ward's, 1956-70). Although national fuel economy did not increase during this period, it declined only slightly from 14.16 to 13.57 miles per gallon or about 4 percent (Federal Highway Administration, 1956-70). This decline is small in view of the fact that manufacturers were building heavier cars and equipping more of them with air conditioners, automatic transmissions, power brakes, power steering.[1]

[1]Between 1956 and 1970 the sales-weighted average curb weight for domestic automobiles increased 7 percent (from 3,280 to 3,520 pounds). In 1956 only 2.8 percent of new car production had factory installed air conditioning; 27.7 percent of new car production had power steering; and 74.8 percent had automatic transmissions. By 1970 these figures had increased to 63.4 percent, 62.5 percent, and 91.1 percent, respectively (Ward's, 1956-70).

VI THE EFFECT OF FEDERAL AUTOMOTIVE EMISSIONS STANDARDS ON VEHICLE FUEL ECONOMY

The regulations of Environmental Protection Agency (EPA) of the U.S. for removing lead additives from gasoline are inseparably related to the Agency's overall strategy for controlling automobile pollution. The foregone energy benefits of removing lead from gasoline must, therefore, be analyzed within the context of automobile emissions standards and the automobile industry's response to meeting those standards.

In 1971 Detroit began decreasing compression ratios to accommodate low-octane, unleaded gasoline in preparation for the introduction of lead sensitive catalytic converters in 1975 and 1976. On a sales-weighted basis, the average compression ratio of all cars dropped from 9.27:1 to 8.65:1 in 1971. Compression ratios were reduced again in 1972 to 8.45:1 and in 1973 to 8.3:1 on a sales-weighted average.

Based on measurements of the fuel economy of new cars, La Pointe (1973) has reported that the 1971 drop in compression ratios has resulted in an accumulated loss in fuel economy of 4.5 - 4.9 percent over uncontrolled cars, or a 3 percent loss compared to 1970 production models. The overall drop in compression ratios from 9.27:1 to 8.3:1 should increase gasoline consumption by 6 - 8 percent.

In addition to losses of efficiency as measured above, there is a study prepared for the American Petroleum Institute (Bonner and Moore, 1967), which estimates that unleaded gasoline would cost 1.8 to 4.7 cents per gallon more if lead were immediately removed and octane maintained. This is relative to a price of 30 cents at the time and represents a 6 to 15% increase.

The results may be summarized with three alternative statistics. First is the increased octane rating of gasoline from use of lead. This increase was 6 to 8%. Second was the increased gasoline consumption required by the reduction of lead currently underway. This increase was 6 to 8%. Finally, we had the increased cost of comparable gasoline without lead which was 6 to 15%. These figures arose from three different approaches to the increased producers' cost of providing automotive transportation units. Through each, the underlying cost structure rises at least 6% and we shall proceed employing this conservative figure.

VII POLICY STRATEGIES AND CONSUMER COST

For simplicity we presume only one grade of gasoline is available. There are two policy alternatives available. One is to allow unlimited lead to be in gasoline. Within this we assume a world without "catalytic converters" so all cars are compatible with leaded gasoline. The alternative is complete removal of lead. Obviously, some intermediate policies may exist. In the long run, technology may well phase out catalytic converters so that even with general pollution control, lead may be reintroduced into gasoline in concentrations which existed prior to the use of catalytic converters.

The supply and demand schedules are shown in Figure 1.

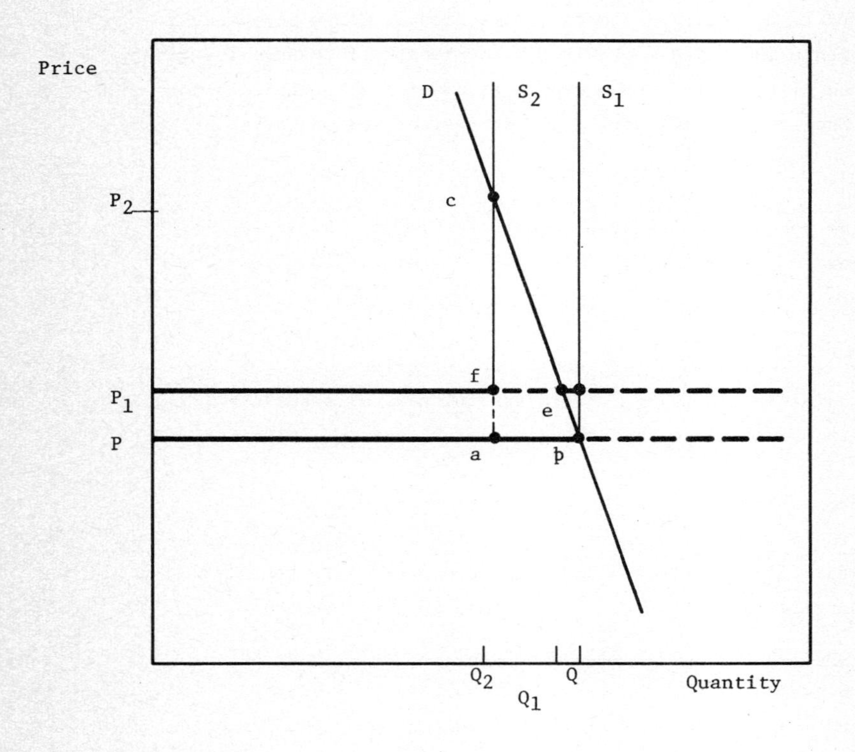

FIGURE 1

SUPPLY AND DEMAND FOR GASOLINE

The initial supply schedule with an import quota is S_1. After lead is removed, supply is S_2. Demand is D. Without import quotas, supply schedules are horizontal lines at P and P_1. If you prefer to consider gasoline as being reduced in octane (quality), then Q measures ultimate miles of travel.

Since costs are presumed to rise 6%, and quantity available reduced by a like amount, $P_1 = 1.06$ P and $Q_2 = Q/1.06$. Demand elasticity is ε. From it one may show $Q_1 = (1 - .06\varepsilon)Q$ and $P_2 = [1 + (1 - 1/1.06)/\varepsilon]P$. Since long run elasticity ε =

.21, Q_1 = .9874Q and P_2 = 1.2696P.

First consider horizontal demand (no import quota). In this case the loss from lead elimination is the area of $PbeP_1$, in Figure 1, which is the added cost of fuel purchases plus lost consumer surplus. This is .05963 PQ.

Next, assume the import quota is invoked but that consumer prices are held down by price control to P_1. (This would prevent excess profits from accruing to oil companies due to the quota.) In fact, this result could be achieved by a price of P_2 coupled with a 100% excess profit tax and rebate of revenue to all consumers. We call this case price control even though it has far more applicability than this name implies. The loss in this case is extra cost of new purchases ($PafP_1$) plus lost consumers surplus (abc).

The extra cost of new purchases are $(P_1 - P)Q_2$ = (.06/1.06)PQ or .-5660 PQ. The lost consumers surplus is $1/2(P_2 - P)$ $(Q - Q_2)$ or .00763 PQ. Total loss is .06423 PQ.

Finally, we assume an import quota without price control. The loss here is as above except prices rise to P_2. Extra cost of new purchases are .25433 PQ. Consumers surplus is identical to that above so total loss is .26196 PQ.

These proportions must all be applied to appropriate price and quantity. Price is taken to be $.50 and quantity for U.S. is 74 billion gallons and for Illinois is 3,785 million gallons. Applying these values we find these annual losses from lead removal based on the 6% increase in cost structure.

Supply Function	Illinois Loss	U.S. Loss
No import quota	$112.8 million	$2206 million
Import quota and price control	$121.5 million	$2377 million
Import quota only	$495.8 million	$9693 million

These results show first the undesirable effects of import quotas. More generally, however, they show that elimination of lead even without import controls will cost consumers of Illinois $118 million annually and of the U.S. over $2.2 billion annually. By any standard this is a very high cost.

VIII SUMMARY

In this study, which is still very preliminary in nature, we have determined dollar value of risk and benefits of removal of lead from gasoline. These are expressed as annual values.

For Illinois we found the cost of lead removal to be $112.8 million under the most favorable assumptions. The risk of continued emissions are $625 thousand for agricultural risk and $22 million for urban dust risk where the latter is very speculative. Hence, these values tend to indicate that the cost of lead removal is substantially greater than risk. For the United States the ratio of cost to risk would be about the same. Certain states such as Florida may have high agricultural risk while others such as Montana will have little.

There remain major factors which could substantially alter these results. First, other agricultural risks could be found in other areas with different soils. Second, it is likely there is an interactive effect between lead and other elements which may substantially increase risk if those become present. Third, the real health risk of lead is unknown and may be very different from that speculated. Finally, the value of human life may be entirely incorrect in our assumption. If one placed a value such as $100,000 per additional year of life, then the results above would be dramatically reversed.

We anticipate further substantial improvements in knowledge concerning the dangers of lead and look forward to the additional research.

REFERENCES

Agriculture Experiment Station, 1967, Soils of Illinois, Bulletin 725, University of Illinois, Urbana, Illinois.

Bonner and Moore Associates, 1967, U.S. Motor Fuel Economics: The Manufacture of Unleaded Gasoline, Volume 1, prepared for the American Petroleum Institute by Bonner and Moore Associates, Inc., Houston, Texas.

Bureau of Mines, 1965-72, Mineral Industry Surveys, "Motor Gasoline," and "Motor Gasoline," U.S. Department of Interior, Bartlesville, Oklahoma.

Federal Highway Administration 1956-1970, Highway Statistics, U. S. Department of Transportation, Washington, D.C.

La Pointe, Clayton, September 10-13, 1973, "Factors Affecting Vehicle Fuel Economy," SAE Paper No. 730791, Society of Automotive Engineers National Fuels and Lubricants Meeting, Milwaukee, Wisconsin.

Metals Task Force, 1974a, Progress Report for an Interdisciplinary Study of Environmental Pollution by Lead and Other Metals, Institute for Environmental Studies, University of Illinois, Urbana, Illinois.

Metals Task Force, 1974b, Research Proposal for the Continuation of an Interdisciplinary Study of Environmental Pollution by Lead and Other Metals, Institute for Environmental Studies, University of Illinois, Urbana, Illinois.

Miller, J.E., Hassett, J., Koeppe, D.E., Rolfe, G.L., and Wheeler, G.L., August 28-31, 1974, "Effects of Soil Properties on Lead Uptake by Corn and Effects of Lead and Cadmium on Corn Root Elongation," Second Annual NSF Trace Contaminants Conference, Asilomar, California.

Ward's, 1956-1970, Ward's Automotive Yearbook, Ward's Communications, Inc., Detroit, Michigan.

COMMENTS ON RESEK-PROVENZANO PAPER[1]

Peter M. Meier

Resek-Provenzano paper raises several interesting questions for energy policy-making, and highlights important methodological problems. The most immediate prerequisite for the application of cost-benefit techniques to public policy analysis is a generally acceptable method of translating qualitative measures of costs (or detrimental impacts) to some quantitative measure;[2] and, as noted by the authors for the lead example, this is considerably easier for such impacts as reduced crop yields than for impacts on human health. Nevertheless, countless recent studies have made the attempt of monetarizing health impacts in areas of policy analysis that are of comparable if not greater significance to the national economy, and also involving trade-offs between economic, environmental and energy goals. Good examples include Sagan's study of health costs in the coal mining industry (of relevance to the debate over fossil versus nuclear power generation);[3] Buehler's monetarization of the value of human life in connection with investments for flood control structures;[4] or more general attempts applying utility theoretical concepts to public polciy analysis.[5]

---

[1]This report was prepared as an account of work sponsored by the United States Government. Neither the United States nor the United States Energy Research and Development Administration, nor any of their employees, nor any of their contractors, subcontractors, or their employees, makes any warranty, express or implied, or assumes any legal liability of responsibility for the accuracy, completeness of usefulness of any information, apparatus, product or process disclosed, or represents that its use would not infringe privately owned rights.

[2]For a good overview of benefit cost analysis and policy making see R.H. Haverman et al., "Benefit-Cost and Policy Analysis," Aldine, Chicago, 1974.

[3]L. A. Sagan, "Health Costs Associated with the Mining Transport and Combustion of Coal in the Steam Electric Industry," Nature 250 (July 12, 1974) 107-111.

[4]B. Buehler, "Monetary Values of Life and Health," Journal, Hydraulics Division, American Society of Civil Engineers 101, No. HY1, (Jan. 1975), 29-47.

[5]J. Hirshleifer, T. Bergstron and E. Rappaport, "Applying Cost-Benefit Concepts to Projects Which Alter Human Morality," University of California at Los Angeles, School of Engineering and Applied Science, Report UCLA-ENG-7478, Nov. 1974. These citations are chosen more as exemplars of alternative analytical approaches than as a complete sample, as the literature in this area has grown significantly over the past few years.

However, such studies have two common characteristics; they tend to be controversial; and, in contrast to the question of automotive lead emissions, the cause and effect relationship is better understood, at least in qualitative terms. Indeed, in the case of lead, the major difficulty is that airborne lead is only one factor, and often a minor one at that, contributing to human lead intake. Except for children ingesting flakes of lead based paint, the largest source of lead exposure is food; the average adult ingests 200-300 micrograms of lead each day, of which 10 percent is absorbed into the bloodstream, with an average lead content of 10 to 30 micrograms per 100 grams of blood.[6] Attempts to establish a direct causal relationship between automotive lead emissions and health impact thus abound with methodological problems, and even EPA acknowledged in 1973 that none of the scientific findings on lead, viewed individually, constituted conclusive evidence that airborne lead by itself was a hazard to human health.[7] However, the EPA position is that when considered together, the studies do indicate that airborne lead is a factor contributing to excessive lead exposure among urban populations. Note that the issue does not center on the health impact of lead, which is beyond dispute, but over the significance of airborne lead vis-a-vis other sources.

In assessing the use of an econometric demand estimation model as a means for deriving the benefits of lead levels in gasoline, it should be remembered that estimates of price and income elasticity vary considerably, depending on their time perspective, geographic focus, and model specification. The short-run price elasticity for the U.S. in a recent Federal Energy Administration forecasting model was taken as -.16;[8] whereas the long-run price elasticity for gasoline in a model of the World Oil Market is estimated by Kennedy as -0.82,[9] with a diversity of intermediate values quoted by other researchers.[10] The problem as far as the United

---

[6] 5 Environmental Law Reporter 2 10052-10056 (1975).

[7] For a good review of the studies quoted by EPA as ground for its intention to impose gasoline lead reductions, see 5 Environmental Law Reporter 20109-20115 (1975).

[8] W. W. Hogan, G. M. Lady and J. D. Pearson, "Petroleum Product Short Term Forecasting at FEA," presented at National Petroleum Refiners Association Computer Conference, San Franciso, Calif., Nov. 1974.

[9] M. Kennedy, "An Economic Model of the World Oil Market," Bell Journal of Economics and Management Science 5, No. 2 (Autumn 1974) 540.

[10] M. S. Houthakker, P. K. Verleger and D. P. Sheehan, in "Dynamic Demand Analyses for Gasoline and Residential Electricity," American Journal of Agricultural Economics (1974), estimate a long-run price elasticity of -0.25.

States is concerned is mainly a statistical one; multi-collinearity in explanatory variables results in high standard error of estimates, and, even if a pooled time series--cross section model of the type used by Resek lessens such difficulty,[11] deficiencies in the data cause other problems of interpretation and specification. Such caveats should be interpreted less as criticisms of Resek's methodology as much as reminders to decision-makers who might use particular numerical estimates in support of a particular policy position.

In regard to other policy matters raised by the paper, it should be noted that the EPA regulations promulgated under the Clean Air Act providing for a phased reduction in the lead content of gasoline are presently before the Court of Appeals in a civil action by the Ethyl Corporation, a major manufacturer of lead additives.[12] In January of this year, the court invalidated the EPA regulations on two grounds; first, that there was insufficient evidence to support a finding that lead additives "endangered" the public health, and that regulations to ban the additives were thus an invalid exercise of administrative discretion;[13] and second, that the administrator made an error in judgement in determining that auto emissions contributed significantly to blood lead levels in adults and children. The court held that the administrator was required to show that air borne lead from auto exhausts contributes a "measurable increment of lead to the human body, and that this measurable increment causes a significant health hazard"; on reviewing the scientific evidence, however, the court found such a conclusion to be unreasonable.[14]

---

[11] Kennedy, see Note 9, supra.

[12] Ethyl Corporation v. EPA (D.C. Circuit, Jan. 28, 1975) 5 Environmental Law Reporter 20096. Other petitioners, whose cases were consolidated into the Ethyl Corporation case for purposes of argument and decision, included PPG Industries, DuPont, NALCO Chemical and the National Petroleum Refiners Association.

[13] The judicial basis for review of an agency action by the courts is the Administrative Procedure Act (5 U.S.C. 701, 1970), which allows judicial reversal or invalidation of an agency action found to be "arbitrary, capricious, and an abuse of discretion" or "unsupported by substantial evidence in a case subject to hearings" see, e.g., D. P. Cume and F. I. Goodman, "Judicial Review of Federal Action: Quest for the Optimum Forum," Columbia Law Review, Vol. 75, No. 1 (Jan. 1975) 1-86.

[14] One should note, however, the distinctions between this case and an earlier case (Amoco Oil Company et al. v. EPA, 501 F 2d 722), in which the same court upheld an EPA regulation that gas stations must provide unleaded gasoline; the evidence that lead destroyed the catalytic converter was uncontested, and therefore fully empowered EPA to take appropriate steps under a provision of the Clean Air Act that would give authority to regulate or control a fuel or fuel additive "...if emission products of such fuel or fuel additive will impair to a significant degree the performance of any emission control device or system which is in general use or which the administrator finds has been developed to a point where in reasonable time it would be in general use," (42 U.S.C. 1857, 1970).

Not unexpectedly, EPA has appealed the 201 decision, and the case will be reheard by the Appeals Court en banc.[15] But regardless of the final outcome, the issue is ultimately whether in such situations the benefit of the doubt should be given to the public health, or to the right of private enterprise to manufacture and sell its products. In particular one should note that the relationship of automotive lead emissions to public health is much more tenuous than in other similar controversies where the evidence for a direct cause-and-effect relationship is more direct--as, for example, in the case of strict new standards governing exposure of workers in the plastics industry to vinyl chloride gas[16]--and this makes the formulation of public policy on the issue a much more complex problem.

In summary, the issue of lead additives in gasoline typifies the complexity of decision-making in energy matters, in which optimal public policy must be based on a very fine balance of competing interests and in the face of considerable uncertainty over the environmental and economic ramifications. Thus research efforts focussed on a clarification of such controversy should continue to be of interest to governmental decision-makers.

---

[15] Cases in the Circuit Appeals Court in the U.S.A. are normally heard by 3 judges; the court may elect, however, to rehear a controversial case en banc, with all justices of that court present.

[16] In that case, the Court of Appeals upheld strict new OSHA (Occupational Safety and Health Administration) standards--see Society of the Plastics Industry v. Occupational Safety and Health Administration, 5 Environmental Law Reporter 20157 (2nd Cir., Jan. 31, 1975).

TECHNICAL PROGRESS IN AGRICULTURE AND ITS IMPLICATIONS FOR ENERGY USE

L. v. Bremen

I. RECENT DEVELOPMENTS OF PRICES, INPUT STRUCTURE AND TECHNICAL PROGRESS

Over the last decade all sectors of industrialized countries have been subject to remarkable factor substitution. This substitution was especially intensive in the field of agriculture. While the number of agricultural workers decreased, the use of other means of production was intensified. But also within these two aggregates of factors mentioned, essential changes of structure have occurred. Changing price relations correspond with the observed factor substitution; this fact is self-evident since the changes in price relations induce substitution. We can see it at rapidly growing wages but a more or less tardy development of other input prices. This situation has changed only recently due to the rapid increase of prices of energy and several raw materials.

Considering the extent of the substitution that has taken place, the development of price relations cannot be the only reason of the changes in input structure. Moreover, technical progress has altered the marginal value products of factors and in this way has induced substitution. This progress, itself, has largely been induced by the evolution of price relations. And at least with regard to the aggregates labor and non-labor inputs, there exists conformity with Hicks' theory of the induction of technical progress, i.e. a higher price of a factor in relation to the prices of other inputs is followed by growing efficiency of this factor in relation to the efficiency of other inputs (Hicks, 1964, pp. 124f). Thus we observe the introduction of labor saving technical progress (Lianos, 1971).

A further characteristic of technical progress in agriculture in the recent past is the reduction of adjustment opportunities. This matter seems to be especially important in view of the subject in question. Until now the opinion is popular that compared to other economic sectors the elasticity of substitution in agriculture is high (Arrow et al., 1961 - Lianos, 1971). But in the case of the industrialized countries this seems to be right only in a modified sense. For agriculture in the Federal Republic of Germany we have found an elasticity of substitution tending towards the rate 2 for the decade 1951-1960 and towards the rate of only 0.6 for the decade 1961-1970. This refers to the elasticity of substitution between the two great aggregates labor and other means of production[1] (without

---

[1] The elasticity of substitution can be written as

(continued on next page)

land which is used on an almost constant level). But we can be sure that in the course of progressive subdivision of the factor side a tendency towards fewer possibilities of substitution can be recognized. Hence, the curvature of isoquants in the agricultural production function has become considerably stronger. Compared to former decades, today a variation in factor price relations causes only a relatively little change in input relations. Farmers using modern technologies of agricultural production have few possibilities of adjustment in the way of varying the input relations by applying less of the comparatively more expensive inputs and more of the comparatively cheaper inputs. Of course, there will be considerable differences between different pairs of inputs. We can observe a relatively high substitution between labor and machines, or between land and fertilizers. On the other hand, only few possibilities of substitution are possible between labor and fertilizers (Hayami et al., 1971, p. 7).

Finally we refer to the enormous energy use in the field of agricultural production as one of the most significant elements of technological development in agriculture. Machines, fuel, fertilizer, and pesticides have been requiring and will require high amounts of energy in their own production and in their use in agriculture. Especially in view of the rapidly growing importance of grain and oilseeds in livestock production a strong increase of energy inputs in food production can be observed. Without regard to the capital using progress in the economic sense we can speak of an energy using progress in the technical sense.

II. AGRICULTURAL ENERGY BALANCES OF VARIOUS REGIONS

In the course of the oil crisis a number of studies dealing with this topic was published. When taking the agricultural sector of the Federal Republic of Germany as an example for the agriculture of a country the characteristics of which are industrial methods of production and, in relation to acreage, intensive use of non-labor inputs, we can make the following statement: In an agricultural sector of the type described the energy input has increased faster than the energy output.

1(continued)

$$\sigma = \frac{\dfrac{d(\frac{C}{L})}{\frac{C}{L}}}{\dfrac{d(\frac{\partial Y}{\partial L} / \frac{\partial Y}{\partial C})}{\frac{\partial Y}{\partial L} / \frac{\partial Y}{\partial L}}}$$

where L = labor
C = non-labor inputs, without land
Y = output

Estimations of σ have been done by regressing the sectoral L/C-ratio on the factor price ratio and trend (Brown, 1968, p. 131).

It is true, the difference between energy input and energy output has become greater in absolute terms. But compared to the growth of gross production this difference has expanded less than proportionately. Corresponding to this fact the input-output relation in energy terms has deteriorated. Behind these general lines of development are important structural changes: On the one hand the energy inputs by draught-animal forage were substituted by those of machine work and those of fuel, lubricants and electrical power. Besides the substitution of energy produced in agriculture itself by fossile energy, this already results in a considerable growth of energy input. The second important item is the increased use of synthetic fertilizer, especially nitrogen. The energy input for seeds shows a declining tendency in the long run which results from the smaller quantities sown per hectar. But here must be conceded that the higher energetic expenditures of hybrid breeding, that cannot be estimated so high in the Federal Republic with its minor corn cultivation as in the United States, have not been taken into account.

Using the same method, we have done calculations for North America based on data published by the Food and Agriculture Organization of the United Nations (FAO). This is a region where agriculture is mechanised. Yet, the factor input per hectar is done with less intensity than in north-western Europe. It comes out clearly that the input-output relation, measured in energy units, is more advantageous in North America than in north-western Europe. Looking at the structure of energy input one can realize that in North America the fertilizer application is as important as in the Federal Republic whereas the use of machines, fuel and electricity is less important. This may have its reason in the differences in intensity mentioned above.

With respect to developing countries such computations are problematic. They can naturally be less representative for real special cases in developing countries than the results mentioned above for industrialized regions. That is why the data used are only averages of utmost divergent single observations. But apart from this, the figures show the less important role of machine work and, resulting from this, low inputs of fuel. On the other hand, high energy inputs from manpower are necessary. The low level of energy input by means of fertilizers shows high reserves of production capacity. But we must not forget that in these regions the possibilities of organic plant nutrition are not fully used, and that in many cases it can be convenient first to make full use of this potential source of plant nutrients (FAO/SIDA 1975). Furthermore, irrigation is not taken into regard in the computations, although irrigation usually is the most important complementary factor of chemical fertilization. Naturally, irrigation itself already is highly efficient. Very often more fuel is needed for this purpose than for powering other machines.

We have not considered the energy input through labor. Since it is immaterial whether people work in agriculture or in another economic sector. They have to be nourished in any case. Secondly, it is rather difficult to compute the energy consumption in human nutrition in units of soil production because of the country-wise very different share of livestock products within people's food basket.

III. ENERGY NEEDS OF LIVESTOCK PRODUCTION AND FOOD PROCESSING

Undoubtedly, at least in the case of the industrialized countries, the computation of agricultural production in energy terms shown above does not show all aspects of the problem. In the industrialized countries and in urbanized regions of developing countries, too, a processing and procurement sector has taken its site between agricultural producers and the consumers of food. This sector on the average works with very high intensity of energy input, although during the recent past its energy using role in relation to that of agriculture seemed not to grow further.

With respect to energy requirements of the total food supply in industrialized countries the recently published studies of Hirst (1974) and of Steinhart/Steinhart (1974), give a better picture, because the authors succeeded in using nearly all energy sources. Their methods of computation do not fully coincide with those Pimentel et al. applied. The most important difference is that Pimentel et al. have considered the example of corn production in the United States while the computations of Hirst and Steinhart/Steinhart refer to the total food sector of the US. Therefore, the computations of the latter result in a considerably less favourable input-output relation in energy terms than our own computations.

The studies of Hirst and Steinhart/Steinhart are essential because they extend the view beyond the limits of the exclusively agricultural production to the total energy needs of food production in an industrialized country with high quality preferences of consumers. They come to the conclusion that 12-13% of the total US-energy consumption are needed for food production. And they show, apart from the energy consumption for processing in the post-agricultural stages, the great influence of agricultural livestock production on energy input in food production. Considering all energy inputs possible, relatively favorable energy ratios are computed for sugar, fats and vegetable oils, flour and cereals, and fresh vegetables. A medium position is occupied by dairy products, eggs, meat and poultry, and fresh fruits. The highest energy ratio is referred in the case of processed fruits and vegetables, and fish. On the other hand, the energy input required per gram of food protein is relatively favorable in the case of fish and very high for fresh fruits, processed fruits and vegetables. In the agriculture of industrialized countries the largest part of soil production is immediately processed, mainly into livestock products. This leads to considerable loss of energy. If the energy problem in its global extension cannot be solved in the meantime, we must expect that, apart from food production and the production of already common raw materials of agricultural origin, the production of fuel for cars, tractors, power stations, and so on, will compete for acreage. A number of scientists have already suggested the absorption of solar energy by means of agriculture for additional purposes. When in the past in several regions of the world the food production was higher than demand for food, this sometimes seemed to be a comfortable outlook into the direction of new markets for agricultural products. But nowadays such aspects imply that in the future people in rich countries will be forced to eat less livestock products and more cereals,potatoes, and so on, only because of decreasing supply of fossil energy in the world (Forthcome, 1968; Delwiche, 1968; Maddox, 1974; Stanford Research Institute, 1973; "Agricultural Situation", 1974).

IV. DIFFERENT EFFECTS OF ENERGY CRISIS IN INDUSTRIALIZED AND DEVELOPING COUNTRIES, POSSIBILITIES OF ADJUSTMENT

Modern technologies of agricultural production are characterized by high levels of energy input of non-agricultural, that is of fossil origin. Econometric analyses do underline facts, that the elasticity of production is very high especially in the case of inputs the production or application of which need great amounts of energy. Particularly in the case of fertilizers the elasticity of production is essentially higher than the average factor share which means a marginal value product also essentially higher than the prices of fertilizers. In developing countries this disequilibrium of factor allocation often is more obvious than in industrialized countries, because of the very different level of application esepcially in the case of fertilizers, but also in the case of pesticides and other inputs of industrial origin. Low levels of application naturally correspond with high marginal values and high elasticities with respect to production (Huang, 1971; Herdt, 1971, v. Bremen, 1974). Because of the high differences between prices and marginal value products in the case of energy intensive inputs, in both regions the demand for these inputs will be inelastic. This results in the well-known high flexibility of prices at world markets which in 1974 could be observed especially in the case of chemical fertilizers.

Apart from the limited probability of less application of the inputs in question which can be explained by economic theory and confirmed by empirical analysis, the question arises whether modern techniques of production do allow considerable adjustment at all. In consequence of the strong curvature of isoquants, respectively the low level of elasticity of substitution, the response curves of single inputs are very steep for the most part. With respect to this, the question arises how much agricultural production would decline if the highly developed technologies failed to get enough of the mentioned essential inputs, especially of such inputs which are produced on basis of hydrocarbons. Naturally, the same question can be asked in the case of other important but less energy intensive raw materials, for instance, phosphates.

From all this we come to different conclusions for industrialized and developing countries. Nowadays, the energy crisis bears more problems for developing countries than for industrialized ones, because of general economic difficulties arising from high prices of crude oil and oil-related products. Many developing countries were not in a position to buy as much fertilizers in 1974 as in preceding years. For these countries the dangers have become clearly visible which arise from the use of modern technologies in the course of "green revolution". In several regions fluctuations in supply with hydrocarbons and its derivatives have led to fluctuations in agricultural production more accentuated than they would have been if conventional technologies of production were used. For instance, in India this resulted in a 4-5 per cent reduction of grain production.

But we must recognize that really serious disturbances in supply with hydrocarbons and other essential raw materials would touch the agricultural production of industrialized countries more than that of developing countries. Modern energy-using technologies are already prevailing in industrialized countries while they are just beginning to spread in developing countries. At all events, if the supply of a country with essential raw materials, especially with hydrocarbons, is endangered, the supply of agriculture and of industries producing important agricultural inputs has to get high priority.

V. OUTLOOK

Generally speaking, the energy crisis has resulted in a world-wide endangering for humanity's food supply; yet this threat is very different in various parts of the world. And we must realize the fact that the deterioration of the political climate having been brought about among others by the oil crisis, has widened this threat to many other raw-material markets. We are surprised to note that there were few people who realized the instability of the situation during the last years. But undoubtedly, today there are already better founded analyses and forecasts than one year ago. In spite of the calm having become greater since then, the situation still is precarious. In any case, in the long run the exhaustion of fossile energy has to be taken into account. For making available totally new sources of energy there still exist only theoretical designs.

Before all, the question arises whether the interrelations between technical progress on the one hand and application of energy on the other may still remain as compulsory as they seemed to be in the past. For, provided that technical trends will remain as they have been up to now, the countries which do not have their own energy resources with respect to their food supply, too, will become more and more dependent on those which export energy. This interdependence could be pushed up to the possibility of physical extortion of overpopulated developing countries, whose possibilities of adjustment by switching from the consumption of highly processed livestock products to less energy requiring foodstuffs of plant origin are fewer than those of the rich industrialized countries.

But as explained above, technical progress in agriculture in a great measure has been induced by price relations. If structures of input prices develop into another direction in the future, should we not expect new sorts of technical progress, too, which could open new paths of adjustment? And could not this new technical progress avoid an important part of the dangers shown above, even if the problem nowadays centrally sited, this is the energy shortage, cannot be solved first at all?

REFERENCES

Arrow, K.J., H.B. Chenery, B.S. Minhas and R.M. Solow, 1961, "Capital-Labor Substitution and Economic Efficiency," The Review of Economics and Statistics, 43, 225-250.

von Bremen, L., 1974, Agrarwirtschaft, 23, 336-347.

Brown, M., 1968, On the Theory and Measurement of Technological Change (Cambridge--University Press).

Delwiche, C.C., 1968, "Nitrogen and Future Food Requirements," in Research for the World Food Crisis, (Ed.) D.G. Aldrich, (Washington, D.C.), pp. 191-210.

Downing, C.G.E. and M. Feldman, 1974, Canadian Farm Economics, 9, No. 1, 24-31.

Food and Agriculture Organization of the United Nations (FAO)/Swedish International Development Authority (SIDA), 1975, "Organic Materials as Fertilizers," Soils Bulletin, 27, (Rome).

Forthcome, P.A., 1968, (FAO Review on Development), Ceres, 1, 50-51.

Hayami, Y., B.B. Miller, W.W. Wade and S. Yamashita, 1971, "An International Comparison of Agricultural Production and Productivities", Technical Bulletin 277-1971, University of Minnesota, USA.

Herdt, R.W., 1971, "Resource Productivity in Indian Agriculture," American Journal of Agricultural Economics, 53, 517-521.

Hicks, J.R., 1964, The Theory of Wages (London--MacMillan).

Hirst, E., 1974, "Food-Related Energy Requirements," Science, 184, 134-138.

Huang, Y., 1971, "Allocation Efficiency in a Developing Agricultural Economy in Malaya," American Journal of Agricultural Economics, 53, 514-516.

Lianos, Th.P., 1971, "The Relative Share of Labor in United States Agriculture," American Journal of Agricultural Economics, 53, 411-422.

Maddox, J., 1974, "Energy and Agriculture," Chemistry and Industry, 124-125.

Perelman, M., 1973, "Mechanization and the Division of Labor in Agriculture," American Journal of Agricultural Economics , 55, 523-526.

Pimentel, D., L.E. Hurd, A.C. Belloti, M.J. Forster, I.N. Oka, O.D. Sholes and R.J. Whitman, 1973, "Food Production and Energy Crisis," Science, 182, 443-449.

Stanford Research Institute, 1973, "Grow Your Own Fuel Supply," News Release, June 11.

Steinhart, J.S. and C.E. Steinhart, 1974, "Energy Use in the U.S. Food System," Science, 184, 307-316.

URBAN DESIGN SHAPING THE ENVIRONMENT - A NOTE

Vladimir Music

Regional and interregional growth of socio-economic activities and consequently of physical assets - usually referred to as the infrastructure, consists of innumerable increments, which ideally would have been consciously shaped and designed. Let us assume that some deliberate form-giving action is present in each process of defining the shell, but also the content of the increment to the growth. Professionals and general public, engaged in this process, and the respective form-giving action might be more or less aware of social, psychological, esthetic, economic and other aspects of an artifact. Therefore we so often speak of well-designed, ill-conceived, functional, dis-functional, beautiful or ugly artifacts or, if you will - interventions into the natural environment. One particular aspect is inviting considerable attention over the last year or two, is the sudden awareness of constraints, imposed by the world-wide energy crisis. In my Yugoslavia, and I can see also elsewhere, steps have been taken to reconsider relative role of various sources of energy in modes of production, transportation, heating etc. Policies have been adopted to impose new limits on travel-speeds, illumination of cities and the like. Prices of fuel, power and consequently of most industrial and agricultural products went soaring up. Up to now, little "institutional" attention has been given to the studies which would explore impacts of higher energy costs on the form of our settlements and regions. Is this partly caused by the fear that some fundamental shifts in the values, social systems, political structure, would be required for implementation of the new solutions, hopefully derived from such studies?

Before we start speculating on alternatives of the urban form in the new conditions, or even of the new spatial relationships in the broader urban system, triggered-off by the impact of energy constraints, we should explore the fundamental attitudes, the society at large holds vis-a-vis urban design, as a conscious and consistent expression of the will of shape the most important part of our built environment, i.e. the urban places. The existing paradigms of urban planning and design process hardly allow for a really broad and at the same time effective (product oriented) participation of the public. In Yugoslavia we are starting to call it - a self-management system of shaping the urban environment - a process in which interests of the planners and the users are supposed to converge. This paper will try to explore the existing and possible new paradigms for the environmental design process, and these shall include as a take-off hypothesis - the self-management participation of citizens, as well as a view that urban design must become an integral part of a more integrated societal planning process. Such requirements might also throw new light on the "intensity of the quality" of design, needed to prevent dilution of the semiotic-symbolic expression of the environmental forms, as an ultimate goal of all cultural endeavours of mankind.

The success of urban/environmental design depends largely on explicitness, as well as on recognition of the existence of all "latent" urban development policies on national and regional levels. It furthermore depends on the product orientation of the general planning process.

As a realistic conept, a diversified set of architectural controls can be envisaged a conceptual device which will take into account the imminent fact that various regional and urban activities are less predictable than some others. The degree of predictability should influence the completeness of design controls, and the openness of systems, provided for the development of the infrastructure. The design criteria, related to such a methodology should include conservation of conventional sources of energy, as well as ways and means of application of new sources, e.g. the solar energy. Such a concept will start from the following bases.

- Current paradigms of the over-all planning process, (economic, social, spatial aspects);
- Position and comprehension of urban/environmental design as a conscious and consistent field of professional activities, ("offensive" and "defensive" attitudes);
- Growth of the self-management socialism as a most viable ideology for the modern society, (self-realization of man vs. modern alienation, associated labor and interest communities, social solidarity, etc.);
- Need for a new system of human values, related to the growth of productive forces, changing socio-economic relations, use of natural resources, balance of materialistic and spiritual spheres, (sciences and arts, technology and culture);
- General evaluation of the role of the supporting "umbrella" sciences: social anthropology, politicology, urbanology, "planology" (studies in planning technology), culturology, semiology, futurology (incl. technological forecasting, especially in energetics and transportation), aesthetics, etc.
- National policies for urbanization, regional and urban development, (explicit ones and "latent" ones);
- The future of urban/environmental design;
- Some operational issues:
  : how to formulate alternatives to the existing (merely spontaneous) regional and urban growth;
  : how to evaluate the alternatives, (some of them may be definite, some merely explorative);
  : the problem of creative authorship as an impediment to new methodologies.

Vol. 115: Anderson et al., Foundations of System Theory: Finitary and Infinitary Conditions. VII, 93 pages. 1976

Vol. 116: K. Miyazawa, Input-Output Analysis and the Structure of Income Distribution. IX, 135 pages. 1976.

Vol. 117: Optimization and Operations Research. Proceedings 1975. Edited by W. Oettli and K. Ritter. IV, 316 pages. 1976.

Vol. 118: Traffic Equilibrium Methods, Proceedings 1974. Edited by M. A. Florian. XXIII, 432 pages. 1976.

Vol. 119: Inflation in Small Countries. Proceedings 1974. Edited by H. Frisch. VI, 356 pages. 1976.

Vol. 120: G. Hasenkamp, Specification and Estimation of Multiple-Output Production Functions. VII, 151 pages. 1976.

Vol. 121: J. W. Cohen, On Regenerative Processes in Queueing Theory. IX, 93 pages. 1976.

Vol. 122: M. S. Bazaraa, and C. M. Shetty, Foundations of Optimization VI. 193 pages. 1976

Vol. 123: Multiple Criteria Decision Making. Kyoto 1975. Edited by M. Zeleny. XXVII, 345 pages. 1976.

Vol. 124: M. J. Todd. The Computation of Fixed Points and Applications. VII, 129 pages. 1976.

Vol. 125: Karl C. Mosler. Optimale Transportnetze. Zur Bestimmung ihres kostengünstigsten Standorts bei gegebener Nachfrage. VI, 142 Seiten. 1976.

Vol. 126: Energy, Regional Science and Public Policy. Energy and Environment I. Proceedings 1975. Edited by M. Chatterji and P. Van Rompuy. VIII, 316 pages. 1976.